LONDON MATHEMATICAL SOCIETY STUDENT TEXTS

Managing editor: Professor C.M. Series, Mathema
University of Warwick, Coventry CV4 7AL, Unite

1 Introduction to combinators and λ-calculus, J.l
2 Building models by games, WILFRID HODGE
3 Local fields, J.W.S. CASSELS
4 An introduction to twistor theory: Second edition, S.A. HUGGETT & K.P. TOD
5 Introduction to general relativity, L.P. HUGHSTON & K.P. TOD
6 Lectures on stochastic analysis: diffusion theory, DANIEL W. STROOCK
7 The theory of evolution and dynamical systems, J. HOFBAUER & K. SIGMUND
8 Summing and nuclear norms in Banach space theory, G.J.O. JAMESON
9 Automorphisms of surfaces after Nielsen and Thurston, A. CASSON & S. BLEILER
10 Nonstandard analysis and its applications, N. CUTLAND (ed)
11 Spacetime and singularities, G. NABER
12 Undergraduate algebraic geometry, MILES REID
13 An introduction to Hankel operators, J.R. PARTINGTON
14 Combinatorial group theory: a topological approach, DANIEL E. COHEN
15 Presentations of groups: Second edition, D.L. JOHNSON
16 An introduction to noncommutative Noetherian rings, K.R. GOODEARL & R.B. WARFIELD, JR.
17 Aspects of quantum field theory in curved spacetime, S.A. FULLING
18 Braids and coverings: selected topics, VAGN LUNDSGAARD HANSEN
19 Steps in commutative algebra, R.Y. SHARP
20 Communication theory, C.M. GOLDIE & R.G.E. PINCH
21 Representations of finite groups of Lie type, FRANÇOIS DIGNE & JEAN MICHEL
22 Designs, graphs, codes, and their links, P.J. CAMERON & J.H. VAN LINT
23 Complex algebraic curves, FRANCES KIRWAN
24 Lectures on elliptic curves, J.W.S. CASSELS
25 Hyperbolic geometry, BIRGER IVERSEN
26 An introduction to the theory of L-functions and Eisenstein series, H. HIDA
27 Hilbert Space: compact operators and the trace theorem, J.R. RETHERFORD
28 Potential theory in the complex plane, T. RANSFORD
29 Undergraduate commutative algebra, M. REID
31 The Laplacian on a Riemannian manifold, S. ROSENBERG
32 Lectures on Lie groups and Lie algebras, R. CARTER, G. SEGAL & I. MACDONALD
33 A primer of algebraic D-modules, S.C. COUTINHO
34 Complex algebraic surfaces, A. BEAUVILLE
35 Young tableaux, W. FULTON
37 A mathematical introduction to wavelets, P. WOJTASZCZYK
38 Harmonic maps, loop groups and integrable systems, M. GUEST
39 Set theory for the working mathematician, K. CIESIELSKI
40 Ergodic theory and dynamical systems, M. POLLICOTT & M. YURI
42 Equilibrium states in ergodic theory, G. KELLER

London Mathematical Society Student Texts 42

Equilibrium States in Ergodic Theory

Gerhard Keller
University of Erlangen

CAMBRIDGE UNIVERSITY PRESS
Cambridge, New York, Melbourne, Madrid, Cape Town, Singapore,
São Paulo, Delhi, Dubai, Tokyo

Cambridge University Press
The Edinburgh Building, Cambridge CB2 8RU, UK

Published in the United States of America by Cambridge University Press, New York

www.cambridge.org
Information on this title: www.cambridge.org/9780521595346

First published 1998

A catalogue record for this publication is available from the British Library

ISBN 978-0-521-59420-2 Hardback
ISBN 978-0-521-59534-6 Paperback

Transferred to digital printing 2010

Contents

Preface

This book grew out of a graduate course on equilibrium states in ergodic theory which I gave twice (in winter 1994/95 and 1996/97) at the University of Erlangen. The text, as it is now, covers rather exactly 32 lectures of 90 minutes each plus a number of exercises. As my intention was (and still is) to put no more into this book than I was able to teach in such a one semester course, the material which I finally included strongly reflects my personal view of the field and is limited in both breadth and depth. But where the knowledgeable reader may wonder why certain topics are omitted or treated only under restrictive assumptions, the student who reads this book will hopefully profit from the concentration on key concepts and key examples gained in this way.

My goal in writing this book was to provide an introduction to the ergodic theory of equilibrium states which gives equal weight to two of its most important applications, namely to equilibrium statistical mechanics and to (time discrete) dynamical systems. In selecting the material, I always kept in mind some of the prime examples of these two fields: the two-dimensional Ising model, piecewise differentiable maps of an interval, and conformal iterated function systems. After working through the book, the reader not only has a solid basic knowledge of the general theory of equilibrium states, but has also seen applications of the theory to such different concepts as phase transitions, observable measures, and the dimension of fractals.

The book starts with a chapter on equilibrium states on finite probability spaces which motivates most of the theory developed later. In this setting, equilibrium states are just probability vectors maximizing a quantity of the type "entropy + energy", i.e., they are distinguished by means of variational principles. Gibbs measures, large deviation estimates, the Ising model on a finite lattice, equilibrium states adapted to Markovian transition mechanisms and absolutely continuous invariant measures are discussed in a nontechnical and completely elementary setting.

In the second chapter, the measure theoretic framework for the general theory is set up: measure preserving $\mathbb{Z}^d$- or $\mathbb{Z}_+^d$-actions on probability spaces. Besides the ergodic theorem the ergodic decomposition is also proved for this class of systems. An extra chapter is devoted to entropy theory for measure preserving actions. It provides a number of tools for entropy calculations needed later, most notably the Shannon–McMillan–Breiman theorem (for $d = 1$) and the Kolmogorov–Sinai theorem.

Chapter 4 on equilibrium states and pressure forms the core of the book. The pressure $p(\phi)$ of a continuous function ϕ is defined from a variational point of view, and the identification of the pressure as a quantity that can be approximated by evaluating ϕ at finitely many points (the "variational principle") is postponed to Section 4.4. The proof of this result is organized in such a way that in certain cases equilibrium states are also constructively approximated. But before this is done, continuity properties of the entropy function and the convex geometric meaning of equilibrium states as derivatives of the pressure function $\phi \mapsto p(\phi)$ are elucidated. Two aspects of this chapter may be particularly noteworthy: (1) Nearly all results apply not only to continuous ϕ but also to upper semicontinuous ones. This allows us e.g., to include interval maps with derivative $+\infty$ in the considerations of the last chapter. (2) The convex geometric point of view makes it possible to prove for expansive actions in Theorem 4.5.9 that the set of all equilibrium states for a given ϕ is the closed convex hull of the set of those equilibrium states that are constructed in the course of the proof of the variational principle.

Chapter 5 is devoted to Gibbs measures. Although this notion has proved very useful for smooth dynamical systems also, we concentrate here on the statistical mechanical setting and restrict to shift spaces over the d-dimensional integer lattice. Gibbs measures for regular functions ϕ are introduced and, using some of the convex geometric results from Chapter 4, they are identified as equilibrium states and vice versa. Furthermore, a large deviations principle for Gibbs measures is proved, stationary measures for finite state Markov chains are identified as Gibbs measures, and the phase transition for the two-dimensional Ising model is investigated.

The last chapter deals with (piecewise) differentiable dynamical systems, in particular with piecewise monotonic interval maps and with iterated function systems. Observability and absolute continuity (w.r.t. Lebesgue measure) of invariant measures are related to their variational properties and simple versions of Rohlin's formula and Ruelle's inequality are derived. For conformal iterated function systems invariant measures of maximal dimension are identified as certain equilibrium states.

The appendix collects a number of facts from analysis, measure theory and probability theory used throughout the book. They are given without proofs but with references to other textbooks.

It is a pleasure for me to thank the following students and colleagues for many helpful remarks and suggestions: Henk Bruin, Bernhard Burgeth, Achim Klenke, Matthias St Pierre, Karl Straußberger, Jan Wenzelburger, Roland Zweimüller. The figures showing equilibrium states of the Ising model at various temperatures were contributed by Achim Klenke. Although I did not follow any existing text, the courses (and hence these notes) were certainly influenced by many books from which I learned ergodic theory. They are all contained in the bibliography.

The index at the end of the book is preceded by a list of special notations which hopefully will help to settle most notational problems. Standard mathematical symbols do not need extra mentioning except perhaps for the conventions that $\mathbb{N} = \mathbb{Z}_+ = \{0, 1, 2, \ldots\}$ and that $O(f(n))$ and $o(f(n))$ may stand for any term $g(n)$ such that $\sup_{n\in\mathbb{N}} \frac{|g(n)|}{|f(n)|} < \infty$ or $\lim_{n\to\infty} \frac{|g(n)|}{|f(n)|} = 0$, respectively.

Erlangen, May 1997,
Gerhard Keller

Chapter 1

Elementary examples of equilibrium states

Equilibrium states, a concept originating from statistical mechanics, are probability measures on topological spaces that are characterized by variational principles. They maximize the sum (or difference) of an entropy and an energy like quantity. Depending on the special choice of the variational problem these measures can have various interesting properties. Some of them are discussed in this introductory chapter on a rather elementary level. Starting from the roots of these ideas in equilibrium statistical mechanics, we outline the connection between equilibrium states and the theory of large deviations and introduce the Ising model of ferromagnetism on a finite lattice as a more concrete example (Sections 1.1 and 1.2). Then we indicate how Markov measures on finite alphabets can be characterized by a variational principle (Section 1.3), and the final section deals with the role played by equilibrium states in the ergodic theory of dynamical systems. That section also furnishes the background for a first discussion of Birkhoff's ergodic theorem.

1.1 Equilibrium states in finite systems

A physical system consisting of many particles can be described on two levels: Microscopically it is determined by its *configuration*, i.e., by the positions and momenta of all particles. Knowing the configuration of a system which obeys the laws of classical mechanics and which is not influenced from outside allows one in principle to determine its exact configuration at any future time. Of course, this fact is of little practical relevance, because the configuration of a realistic large system (e.g., the positions and momenta of all $2.7 \cdot 10^{22}$ molecules of an ideal gas in a one litre container) cannot even approximately be known. On the other hand, a good description of the macroscopic properties of such a system is provided by a relatively small number of observable parameters like total energy, temperature, entropy, etc. Mathematically, these macroscopic quantities are parameters associated with probability distributions on the space of all configurations. We call such a distribution a *state*.

This point of view was developed towards the end of the 19th century in the works of Ludwig Boltzmann (1844–1906) and John Willard Gibbs (1839–1903) in their attempt to reconcile the irreversibility of (macroscopic) thermodynamic processes with the reversibility of the supposed underlying mechanical motions. A historical document that summarizes the debates of this period with great clarity is the article [18] by P. and T. Ehrenfest. Even today it is a pleasure to read this account of the pioneering ideas of Boltzmann and Gibbs that, although written at a time where measure theory and modern probability theory were not yet invented, describes exactly the mathematical shortcomings of the theory at that time. Today, nearly one century later, a broad mathematical discipline called *ergodic theory* provides a strong foundation for a better understanding of (not only) these ideas.

If a physical system that is confined to a finite volume is not influenced from outside, it is driven by its internal fluctuations towards an *equilibrium state*, i.e., it prefers configurations which are compatible with the macroscopic parameters of this state (and thus do not reveal any additional useful information about the system). In order to make these very vague considerations more precise, let us consider an elementary probabilistic model that reflects the two "Fundamental Theorems of Thermodynamics":

1. The energy of a closed system is constant.
2. The entropy of such a system is maximized by its equilibrium states.

In this section we consider an elementary probabilistic setting that allows us to define quantities called entropy and energy which are associated with probability measures, and we describe those measures that maximize the entropy under the constraint of keeping the energy constant. Despite its simplicity this setting already contains germs of many of the technically more difficult definitions and proofs we shall meet in later chapters.

Let Ω be a finite set, our (abstract) configuration space.

▷ A *state* is a probability vector $\mu = (\mu(\omega) \mid \omega \in \Omega)$. The set of all states is denoted by $\mathcal{M}$.

▷ The *entropy* of the state μ is defined as $H(\mu) := -\sum_{\omega\in\Omega} \mu(\omega) \log \mu(\omega)$. It is continuous as a function of $\mu \in \mathcal{M} \subset \mathbb{R}^{\Omega}$.

1.1.1 Remark At this point the reader should not try to interpret the expression for $H(\mu)$ as the classical thermodynamic entropy. Instead, we take it as a measure for the amount of uncertainty that the observer is left with when he knows that the system is in state μ. The connection between the information theoretic and the thermodynamic aspects of entropy is discussed later.

Let $A \subseteq \Omega$. With the event "$\omega \in A$" we want to associate a positive real number I_A that can be interpreted as its amount of information. If one requires that I_A is a continuous function of the probability $\mu(A) := \sum_{\omega\in A} \mu(\omega)$ only and that $I_{A\cap B} = I_A + I_B$ for statistically independent events A and B, the only possible choice is $I_A = -\log\mu(A)$ where the logarithm can be taken to any base. For convenience we shall always use the natural logarithm. Now $H(\mu) = \sum_{\omega\in\Omega} \mu(\omega) I_\omega$ is just the average amount of information of the elementary events ω. In other words, $H(\mu)$ is the expected amount of information that can be gained from further observations on the system, if its present state is known to be μ. That the maximal value of $H(\mu)$ is $\log|\Omega|$ can be seen as follows:

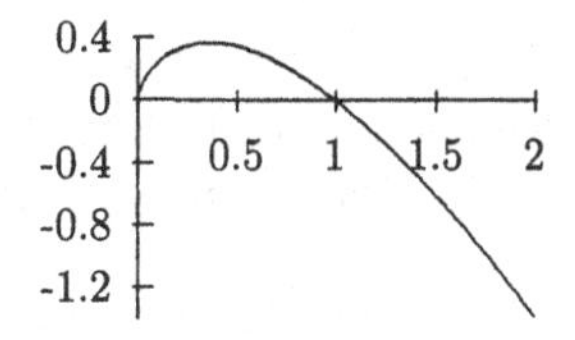

Figure 1.1: The function φ

Let $\varphi : [0,1] \to [0,\infty)$, $\varphi(t) = -t\log t$, see Figure 1.1. φ is continuous and concave and $H(\mu) = \sum_\omega \varphi(\mu(\omega))$. Throughout this book, φ will always denote this function.

Set $n := |\Omega|$. By an elementary version of Jensen's inequality (see A.2.2),

$$\frac{1}{n}\sum_\omega \varphi(\mu(\omega)) \le \varphi\left(\frac{1}{n}\sum_\omega \mu(\omega)\right) = \varphi\left(\frac{1}{n}\right)$$

so that

$$H(\mu) = \sum_\omega \varphi(\mu(\omega)) \le n\cdot\varphi\left(\frac{1}{n}\right) = H\left(\left(\frac{1}{n},\dots,\frac{1}{n}\right)\right) = \log n\,.$$

Let μ and μ' be probability vectors on spaces Ω and Ω' respectively. The additivity of the information $I_{A\cap B} = I_A + I_B$ for independent events implies

$$H(\mu\times\mu') = H(\mu) + H(\mu')\,.$$

◇

We continue specifying our model:

- ▷ Each configuration $\omega \in \Omega$ is assigned an energy value $u(\omega) \in \mathbb{R}$ such that in state μ the system has mean energy $\mu(u) := \sum_{\omega\in\Omega}\mu(\omega)u(\omega)$. By $V_\mu(u) := \mu(u^2) - (\mu(u))^2$ we denote the variance of u under μ.
- ▷ $Z(\beta) := \sum_{\omega\in\Omega}\exp(-\beta u(\omega))$ is the *partition function* of u. Although it is defined for complex β, we are mostly interested in its values for real arguments.

▷ For real parameters β the *Gibbs measure* μ_β on Ω is defined by

$$\mu_\beta(\omega) := \frac{1}{Z(\beta)} \exp(-\beta u(\omega)) .$$

As $\frac{\mu_\beta(\omega)}{\mu_\beta(\omega')} \to 0$ for $\beta \to +\infty$ if $u(\omega) > u(\omega')$, the measures μ_β converge to the equidistribution on $\Omega_{min} := \{\omega : u(\omega) = \min_\Omega u\}$ if $\beta \to +\infty$. Since an analogous argument holds for the limit $\beta \to -\infty$, it follows in particular that

$$\lim_{\beta\to+\infty} \mu_\beta(u) = \min_\omega u(\omega) \quad \text{and} \quad \lim_{\beta\to-\infty} \mu_\beta(u) = \max_\omega u(\omega) . \tag{1.1}$$

1.1.2 Lemma *$\beta \mapsto \log Z(\beta)$ is a real analytic map and*

$$(\log Z)'(\beta) = -\mu_\beta(u) \quad \text{and} \quad (\log Z)''(\beta) = V_{\mu_\beta}(u) \geq 0$$

with equality if and only if the function u is constant. In particular, $\beta \mapsto \log Z(\beta)$ is convex.

Proof: $Z(\beta) = \sum_{\omega\in\Omega} \exp(-\beta u(\omega))$ is analytic and nonzero for $\beta \in \mathbb{R}$ such that $\log Z(\beta)$ is real analytic. Elementary differentiation yields

$$(\log Z)'(\beta) = \frac{Z'(\beta)}{Z(\beta)} = -\frac{1}{Z(\beta)} \sum_{\omega\in\Omega} u(\omega) e^{-\beta u(\omega)} = -\mu_\beta(u)$$

and

$$\begin{aligned}(\log Z)''(\beta) &= \frac{Z''(\beta)}{Z(\beta)} - \left(\frac{Z'(\beta)}{Z(\beta)}\right)^2 = \frac{1}{Z(\beta)} \sum_{\omega\in\Omega} u^2(\omega) e^{-\beta u(\omega)} - (\mu_\beta(u))^2 \\ &= V_{\mu_\beta}(u) .\end{aligned}$$

□

The following theorem is an elementary prototype of a much more general statement that is proved in a later section and that characterizes Gibbs measures by means of a variational principle.

1.1.3 Theorem (Variational principle)
Each Gibbs measure μ_β with $\beta \in \mathbb{R}$ satisfies

$$H(\mu_\beta) + \mu_\beta(-\beta u) = \log Z(\beta) = \sup_{\nu\in\mathcal{M}} (H(\nu) + \nu(-\beta u)) . \tag{1.2}$$

A measure ν for which this supremum is attained is called an equilibrium state for $-\beta u$. Thus Gibbs measures are equilibrium states. In fact, μ_β is the only equilibrium state for $-\beta u$.

Proof: Of course this theorem can be proved by elementary calculus using Lagrange multipliers. We prefer to give a proof based on a convexity argument (Jensen inequality, see A.2.2) applied to the concave function $x \mapsto \log x$. For $\nu \in \mathcal{M}$ we have

$$\begin{aligned} H(\nu) + \nu(-\beta u) &= -\sum_{\omega\in\Omega} \nu(\omega)(\log \nu(\omega) + \beta u(\omega)) \\ &= \sum_{\omega\in\Omega} \nu(\omega) \log \frac{e^{-\beta u(\omega)}}{\nu(\omega)} \\ &\leq \log \sum_{\omega\in\Omega} \nu(\omega) \frac{e^{-\beta u(\omega)}}{\nu(\omega)} \\ &= \log Z(\beta) \end{aligned}$$

with equality if and only if the random variable $\omega \mapsto \frac{\exp(-\beta u(\omega))}{\nu(\omega)}$ is constant, i.e., if $\nu = \mu_\beta$. □

1.1.4 Remark Although *equilibrium* is rather a dynamical notion that can hardly be separated from the idea of a dynamical process evolving towards it (in the case of a stable equilibrium) or away from it (if the equilibrium is unstable), the variational principle makes no statement about the time evolution of non-equilibrium initial states. It just characterizes the equilibrium on the basis of physical principles that are beyond the scope of this text. ◇

1.1.5 Corollary *If E^* is a real number and if* $\min_\omega u(\omega) < E^* < \max_\omega u(\omega)$, *then there is a unique parameter $\beta^* \in \mathbb{R}$ such that μ_{β^*} has energy $\mu_{\beta^*}(u) = E^*$ and maximizes entropy among all states ν with the same energy E^*.*

Proof: It follows from Lemma 1.1.2 that $\beta \mapsto \mu_\beta(u)$ is a continuous and strictly decreasing function of β. As $\lim_{\beta\to-\infty} \mu_\beta(u) = \max_\omega u(\omega)$ and $\lim_{\beta\to+\infty} \mu_\beta(u) = \min_\omega u(\omega)$ by (1.1), there is a unique parameter β^* with $\mu_{\beta^*}(u) = E^*$. Hence, for any state ν with $\nu(u) = E^*$ the variational principle implies

$$H(\nu) - \beta^* E^* = H(\nu) + \nu(-\beta^* u) \leq H(\mu_{\beta^*}) + \mu_{\beta^*}(-\beta^* u) = H(\mu_{\beta^*}) - \beta^* E^* ,$$

i.e., $H(\nu) \leq H(\mu_{\beta^*})$, and equality holds if and only if $\nu = \mu_{\beta^*}$. □

1.1.6 Remark In a physical context, $T = 1/\beta$ denotes the temperature of a system and $\mu_{1/T}$ describes the equilibrium of the system at temperature T. (For simplicity we set the Boltzmann constant $k = 1$.) Using the notation $F(\nu) := \nu(u) - T \cdot H(\nu)$ for the *free energy* of the state ν, we can reformulate the variational principle in the following way:

$$F(\nu) \geq F(\mu_{1/T}) = -T \log Z(1/T) \text{ with equality if and only if } \nu = \mu_{1/T}. \quad (1.3)$$

Even more can be said: In the next lemma we prove $\frac{d}{d\beta}H(\mu_\beta) = \beta \cdot \frac{d}{d\beta}\mu_\beta(u)$. It follows immediately that

$$\frac{d}{dT}H(\mu_{1/T}) = \frac{1}{T} \cdot \frac{d}{dT}\mu_{1/T}(u)$$

where $\mu_{1/T}(u)$ is the energy of the system at temperature T. Compare this to the usual thermodynamic definition of the entropy S of a system by its differential $dS = \frac{dQ}{T}$, where Q denotes the heat content of the system and T is its absolute temperature. ◇

1.1.7 Lemma

$$\frac{d}{d\beta}H(\mu_\beta) = \beta \cdot \frac{d}{d\beta}\mu_\beta(u) .$$

Proof: Recall from the variational principle that $H(\mu_\beta) = \log Z(\beta) + \beta\mu_\beta(u)$. By Lemma 1.1.2 we have $\frac{d}{d\beta}\log Z(\beta) = -\mu_\beta(u)$. Hence

$$\frac{d}{d\beta}H(\mu_\beta) = -\mu_\beta(u) + \mu_\beta(u) + \beta\frac{d}{d\beta}\mu_\beta(u) = \beta\frac{d}{d\beta}\mu_\beta(u) .$$

□

1.1.8 Remark Recall that $\mu_\beta(\omega) = \frac{1}{Z(\beta)}e^{-\beta u(\omega)}$. For high temperatures $T \to +\infty$ (i.e., $\beta \searrow 0$) the equilibrium states μ_β converge to the *equidistribution* on Ω, which is the state maximizing the entropy (i.e., maximizing the disorder of the system). At low temperatures $T \searrow 0$ (i.e., $\beta \to +\infty$) the μ_β converge towards the *ground state*, i.e., towards the equidistribution on $\Omega_{min} := \{\omega : u(\omega) = \min_\Omega u\}$, see (1.1). ◇

1.1.9 Remark In later sections we mostly fix $-\beta$ and include it in the function u. In other words, we study the case $\beta = -1$ for which (1.2) becomes

$$p(u) := \sup_{\nu \in \mathcal{M}} (H(\nu) + \nu(u)) = \log Z(-1) = \log \sum_{\omega\in\Omega} e^{u(\omega)}$$

where the supremum is attained if and only if $\nu(\omega) = e^{u(\omega)-p(u)}$. ◇

1.1.10 Exercise By a slight modification of the proof of Theorem 1.1.3 show that for each $\nu \in \mathcal{M}$

$$H(\nu) = \inf_{u:\Omega\to\mathbb{R}} (p(u) - \nu(u)) .$$

◇

1.2 Systems on finite lattices

In more realistic physical models configurations usually have some spatial structure. Mathematically this means that there is a group of isometries acting on an underlying position space. We consider the following elementary example of such a situation:

- ▷ The configuration space is of the form $\Omega = \Sigma^G$, where the additive group $G = (\mathbb{Z}/\ell\mathbb{Z})^d$ is geometrically interpreted as a finite grid, and Σ is the finite set of possible values of a configuration at a given site. Elements $\omega \in \Omega$ are written as $\omega = (\omega_g)_{g\in G}$.
- ▷ The group G operates on Ω by an action $\mathcal{T} = (T^g : g \in G)$ where the maps $T^g : \Omega \to \Omega$ act as *shift* transformations on Ω, namely $(T^g\omega)_i := \omega_{i+g}$.
- ▷ $\mathcal{M}(\mathcal{T})$ denotes the set of all probability vectors ν on Ω that are invariant under $\mathcal{T}$, i.e., for which $\nu(T^{-g}\omega) = \nu(\omega)$ for all $g \in G$ and all $\omega \in \Omega$. If $\nu \in \mathcal{M}(\mathcal{T})$, then $\nu(u \circ T^g) = \sum_\omega \nu(\omega)u(T^g\omega) = \sum_\omega \nu(T^{-g}\omega)u(\omega) = \sum_\omega \nu(\omega)u(\omega) = \nu(u)$.
- ▷ The energy function u has the form $u(\omega) = \sum_{g\in G}\psi(T^g\omega)$ with a *local energy function* $\psi : \Omega \to \mathbb{R}$. In this case $u(T^g\omega) = u(\omega)$ for all $g \in G$. As $\mu_\beta(\omega) = \frac{1}{Z(\beta)}e^{-\beta u(\omega)}$, it follows that $\mu_\beta \in \mathcal{M}(\mathcal{T})$ (i.e., the Gibbs distribution is spatially homogeneous). In particular $\nu(u) = |G|\,\nu(\psi)$ for each $\nu \in \mathcal{M}(\mathcal{T})$.
- ▷ Define the *pressure* of $-\beta\psi$ as

$$p(-\beta\psi) := \sup_{\nu\in\mathcal{M}(T)}\left(\frac{1}{|G|}H(\nu) + \nu(-\beta\psi)\right).$$

Then the variational principle can be written as

$$\frac{1}{|G|}H(\mu_\beta) + \mu_\beta(-\beta\psi) = \frac{1}{|G|}\log Z(\beta) = p(-\beta\psi)\,. \tag{1.4}$$

From a physical point of view the word *pressure* should not be taken too literally, because

$$p(-\beta\psi) = -\frac{1}{T}\,\frac{F(\mu_{1/T})}{|G|}$$

in terms of the free energy $F(\nu) = \nu(u) - TH(\nu)$, whereas in a thermodynamic context $p = -\frac{\partial F}{\partial V}$ where V denotes the volume of the system and p its pressure. Therefore it might be more appropriate to call $\frac{1}{\beta}p(-\beta\psi)$ the pressure (this is $p(-\psi)$ if $\beta = T = 1$), but this is not the common usage in ergodic theory.

Note also that the term $\frac{1}{|G|}H(\mu_\beta)$ entering the definition of pressure can be interpreted as the entropy per lattice site of the measure μ_β.

1.2.1 Example (Bernoulli measures and large deviations) Suppose that the local energy depends on a single lattice site (which can be assumed without loss of generality to be $g = 0$). So $\psi(\omega) = f(\omega_0)$ for some $f : \Sigma \to \mathbb{R}$. For $\sigma \in \Sigma$ let

$$q_\beta(\sigma) := \frac{e^{-\beta f(\sigma)}}{\sum_{\tau \in \Sigma} e^{-\beta f(\tau)}}$$

and write $N := |G|$. Then $(q_\beta(\sigma) \mid \sigma \in \Sigma)$ is a Gibbs measure on Σ with $\rho_\beta := \max_\sigma q_\beta(\sigma) < 1$. As $Z(\beta)$ takes the form $Z(\beta) = (\sum_{\tau \in \Sigma} e^{-\beta f(\tau)})^N$, we have for each $\omega \in \Omega$

$$\mu_\beta(\omega) = \frac{1}{Z(\beta)} \exp\Big(-\beta \sum_{g \in G} f(\omega_g)\Big) = \prod_{g \in G} q_\beta(\omega_g) ,$$

i.e., $\mu_\beta = q_\beta^{\times G}$ is a Bernoulli measure. In particular, $\mu_\beta(\omega) \leq \rho_\beta^N$. Hence, the probability $\mu_\beta(\omega)$ of each configuration is exponentially small in the system size N, and it does not make any sense to think of the actually observed configuration of a system in state μ_β as the most probable one. Instead, following an idea of Boltzmann's, we shall argue that μ_β strongly favours such configurations for which the *empirical distribution* π_ω is very close to the one-dimensional marginal distribution of μ_β. Here π_ω is the probability distribution on Σ defined by $\pi_\omega(\sigma) = N^{-1} \operatorname{card}\{g \in G : \omega_g = \sigma\}$. (Section 12b of the Ehrenfests' article [18] is an excellent account of this argument.)

As q_β is the Gibbs measure for $-\beta f$ on Σ, we have

> $H(q_\beta) - \beta\, q_\beta(f) \geq H(\pi) - \beta\, \pi(f)$ for each probability vector π on Σ with equality if and only if $\pi = q_\beta$.

So we may measure the distance from a probability vector π to q_β by

$$d_\beta(\pi) := (H(q_\beta) - \beta\, q_\beta(f)) - (H(\pi) - \beta\, \pi(f))$$

and ask for the probability under μ_β of the event $U_{\beta,r} := \{\omega \in \Omega : d_\beta(\pi_\omega) < r\}$. (Geometrically this is the event that π_ω belongs to a convex neighbourhood of q_β in the simplex of probability vectors on Σ, because $d_\beta(\pi)$ is a convex function of π. See Theorem 3.3.2 for the convexity and Figure 1.2 for an illustration in the case $|\Sigma| = 3$.) The result we are going to prove is: if $0 < r < H(q_\beta) - \beta\, q_\beta(f)$, then

$$\lim_{N \to \infty} \frac{1}{N} \log \mu_\beta(\Omega \setminus U_{\beta,r}) = -r . \tag{1.5}$$

This is a simple example of the *large deviations property* of Gibbs measures. It says that under the probability distribution μ_β roughly a fraction e^{-rN} of all configurations does not belong to the neighbourhood $U_{\beta,r}$ of μ_β.

Figure 1.2: Gibbs measures q_β and a convex neighbourhood $\{\pi : d_\beta(\pi) < 0.05\}$ with $\Sigma = \{1, 2, 3\}$, $f(1) = 1, f(2) = 2, f(3) = 3$ and $\beta = 0.5$.

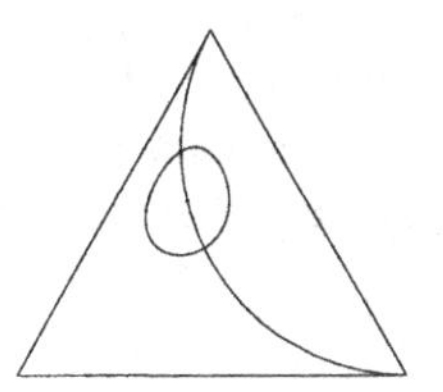

In order to evaluate $\mu_\beta(U_{\beta,r})$ observe that $H(\mu_\beta) = N\,H(q_\beta)$ and therefore

$$\begin{aligned}
\log \mu_\beta(\omega) &= -\beta \sum_{g\in G} f(\omega_g) - \log Z(\beta) \\
&= -N\beta\pi_\omega(f) - (H(\mu_\beta) - N\beta\mu_\beta(\psi)) \\
&= -N \cdot \Big((H(q_\beta) - \beta q_\beta(f)) - (H(\pi_\omega) - \beta\pi_\omega(f))\Big) - N \cdot H(\pi_\omega) \\
&= -N \cdot d_\beta(\pi_\omega) - N \cdot H(\pi_\omega)\,. \qquad (1.6)
\end{aligned}$$

Let $K := \{k \in \mathbb{N}^\Sigma : \sum_{\sigma\in\Sigma} k(\sigma) = N\}$. Then each π_ω is of the form $\pi_\omega = k\,N^{-1}$ for some $k \in K$. Let

$$N(k) := \operatorname{card}\{\omega \in \Omega : \pi_\omega = k\,N^{-1}\} = \frac{N!}{\prod_{\sigma\in\Sigma} k(\sigma)!}\,.$$

We approximate the numbers $N(k)$ by Stirling's formula $n! = \sqrt{2\pi n}\,e^{-n}\,n^n\,e^{r_n}$ where $\lim_{n\to\infty} r_n = 0$. Then

$$N(k) = \text{const} \cdot N^{-\frac{1}{2}(|\Sigma|-1)} \cdot \prod_{\substack{\sigma\in\Sigma \\ k(\sigma)\neq 0}} (k(\sigma)\,N^{-1})^{-k(\sigma)-\frac{1}{2}} \cdot \gamma(N, k)$$

where $\gamma(N, k)$ is bounded away from 0 and ∞ uniformly in N and k. Hence

$$\log N(k) = N \cdot H(k\,N^{-1}) + O(\log N)\,, \qquad (1.7)$$

and we can estimate

$$\begin{aligned}
\frac{1}{N}\log \mu_\beta(\Omega \setminus U_{\beta,r}) &= \frac{1}{N}\log \sum_{\omega\in\Omega\setminus U_{\beta,r}} \mu_\beta(\omega) \\
&\leq \frac{1}{N}\log \sum_{\omega\in\Omega\setminus U_{\beta,r}} \exp(-Nr - NH(\pi_\omega)) \quad \text{by (1.6)} \\
&\leq \frac{1}{N}\log \sum_{k\in K} N(k) \cdot \exp\left(-Nr - NH(k\,N^{-1})\right) \\
&\leq \frac{1}{N}\log\left(|K| \cdot \exp(-Nr + O(\log N))\right) \\
&= -r + O\left(\frac{\log N}{N}\right) \quad \text{as } |K| \leq N^{|\Sigma|},
\end{aligned}$$

i.e., $1-\mu_\beta(U_{\beta,r})$ is exponentially small in the lattice size N. This shows the "$\leq$"-direction of equality (1.5).

On the other hand, let $r'>r$. As $r<H(q_\beta)-\beta\, q_\beta(f)$, there is, for sufficiently large N, some $k\in K$ such that $d_\beta(k\,N^{-1})\in[r,r']$. Therefore (1.6) and (1.7) lead to

$$\begin{aligned}\frac{1}{N}\log\mu_\beta(\Omega\setminus U_{\beta,r}) &\geq \frac{1}{N}\log\sum_{\substack{\omega\in\Omega\\ \pi_\omega=k\,N^{-1}}}\mu_\beta(\omega)\\ &\geq \frac{1}{N}\log\left(N(k)\cdot\exp(-Nr'-NH(k\,N^{-1}))\right)\\ &= -r'+O\left(\frac{\log N}{N}\right).\end{aligned}$$

As $r'>r$ is arbitrary, this finishes the proof of (1.5). ◇

1.2.2 Example (Ising model) A classical example where the local energy ψ depends not only on one lattice site but on local configurations involving several sites is the *Ising model.* It was designed by the physicist Lenz around 1920 to explain ferromagnetism and was named after his student Ising who contributed to its theory. The idea is that iron atoms are situated at the sites of a lattice G and behave like small magnets that may be oriented upwards (denoted by $+1$) or downwards (denoted by -1). Physically, two magnets that are close to one another need less energy to be oriented in the same sense than in an opposite sense. This leads to the following simplified model: $\Sigma=\{-1,+1\}$ represents the possible orientations of the magnetization, $\psi(\omega)=-\sum_{j=1}^{d}(\omega_0\omega_{e_j}+\omega_0\omega_{-e_j})$ is the local energy, where $e_1,\dots,e_d$ denote the canonical unit vectors in G. At positive values of β (i.e., at positive temperatures) $u(\omega)$ is small for rather homogeneous configurations ω. Indeed, it is minimal if ω is constant $+1$ or -1. The *ground state* (also frozen state) is $\lim_{\beta\to+\infty}\mu_\beta=\frac{1}{2}(\delta_{(+1)^G}+\delta_{(-1)^G})$. Here δ_ω denotes the unit point mass in $\omega\in\Omega$, and $(+1)^G$ and $(-1)^G$ are the constant configurations consisting of $+1$'s (respectively -1's) only.

A rather nontechnical introduction to the Ising model is presented in [49, Chapter 1]. We come back to this model in Section 5.5. Figure 1.3 gives an impression of typical Ising configurations. ◇

1.2.3 Exercise Determine $\lim_{\beta\to-\infty}\mu_\beta$ for the Ising model. (This is the ground state of the so called anti-ferromagnetic Ising model which is obtained if the function ψ in the Ising model is replaced by $-\psi$.) ◇

Statistical properties of the equidistribution ($\beta=0$) and quite often also of the ground state ($\beta=\infty$) are rather easy to describe. This is no longer true for

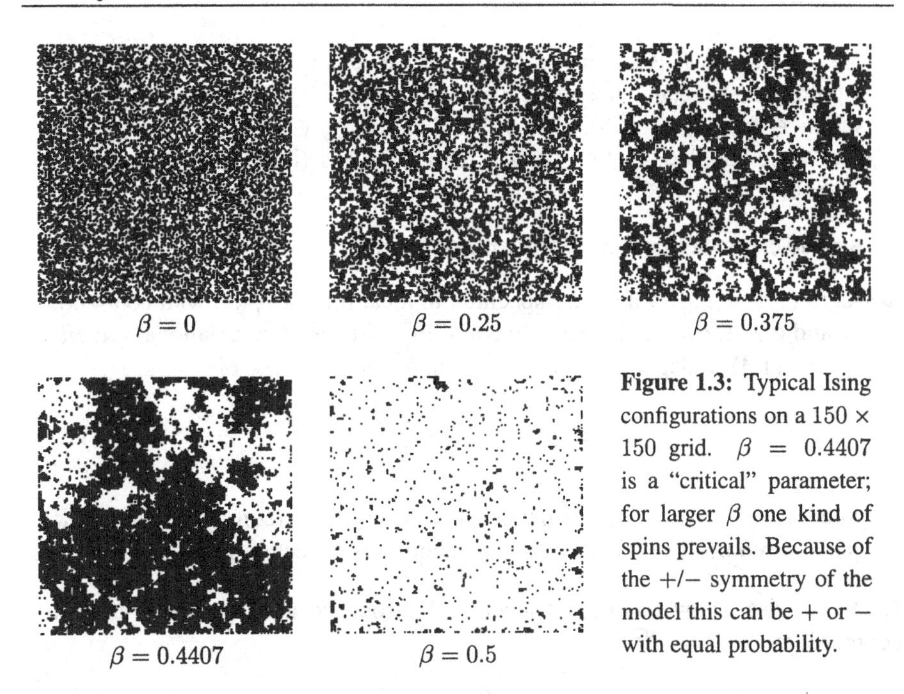

Figure 1.3: Typical Ising configurations on a 150×150 grid. $\beta = 0.4407$ is a "critical" parameter; for larger β one kind of spins prevails. Because of the +/− symmetry of the model this can be + or − with equal probability.

$\beta \in (0, \infty)$ if the lattice size is large. One might ask, for example, whether in the Ising model the Gibbs measure μ_β for large but finite β decomposes – just like the ground state – into two phases, one with dominating $+1$'s and one with dominating -1's as Figure 1.3 for $\beta = 0.5$ suggests. In finite models this question has no meaningful answer, because on a finite lattice each state which is invariant under the action of the group G can be represented as a weighted sum of equidistributions on *group orbits* $G\omega = \{T^g\omega : g \in G\}$ $(\omega \in \Omega)$, and if β is finite, then $\mu_\beta(G\omega) > 0$ for each $\omega \in \Omega$. On the other hand, as demonstrated in a simple setting at the end of Example 1.2.1, most of the mass of μ_β is concentrated on typical configurations, and this effect becomes the more obvious the larger the lattice size is (this is the large deviations property). Therefore it is not too surprising that a satisfactory theory of (non)decomposability of equilibrium states requires systems on infinite lattices $G = \mathbb{Z}^d$, and in fact, we will find out in Section 5.5 that in dimension $d = 1$ there is no "phase decomposition" for the Ising model, i.e., there is always a unique equilibrium state, whereas in dimension $d \geq 2$ there are (at least) two mutually singular equilibrium states for sufficiently large β. This is interpreted as a phase transition. (For small enough β the equilibrium is unique in higher dimensions also.)

This motivates the following setting:

▷ $G = \mathbb{Z}^d$ or $G = \mathbb{Z}^d_+$ is an infinite lattice with corresponding configuration space $\Omega = \Sigma^G$. Endowed with the product topology of the discrete topology on Σ the space Ω is a compact topological space (Tychonov's theorem). This topology is also generated by the metric $d(\omega,\omega') := 2^{-n(\omega,\omega')}$ where $n(\omega,\omega') := \sup\{n \in \mathbb{N} : \omega_g = \omega'_g \text{ for } g \in \Lambda_n\}$ and $\Lambda_n := \{g \in G : |g_1|,\dots,|g_d| < n\}$ is a "cube" of side length n (if $G = \mathbb{Z}_+^d$) or $2n-1$ (if $G = \mathbb{Z}^d$). In particular $\Lambda_0 = \emptyset$.

▷ A *state* is now a Borel probability measure on Ω.

▷ As in (1.4) energy values are specified in local form as $\psi(\omega)$ with continuous ψ, and one aim is to characterize equilibrium states by a variational principle of type (1.4). Independently one may ask for the existence of

$$u(\omega) = \lim_{n\to\infty} \frac{1}{|\Lambda_n|} \sum_{g\in\Lambda_n} \psi(T^g\omega) \tag{1.8}$$

where $(T^g\omega)_i = (\omega)_{i+g}$ as in the finite G setting. In a certain sense this question is answered positively by the ergodic theorem (see Section 2.1).

1.2.4 Exercise Show that the metric $d(\omega,\omega')$ introduced above generates the product topology on $\Omega = \Sigma^G$. ◇

1.3 Invariant distributions for Markov matrices

In this section Σ is again a finite abstract configuration space without any spatial structure. We add a dynamical aspect to our model by introducing a stochastic $\Sigma \times \Sigma$ matrix $Q = (q_{ij})$ which we interpret as a prescription of a Markovian dynamic on Σ. More precisely, q_{ij} is the probability that our system, given that it is in configuration i, goes to configuration j in one unit of time. (So $q_{ij} \geq 0$ for all i, j and $\sum_j q_{ij} = 1$ for all i.) The matrix Q acts by right multiplication $\nu \mapsto \nu Q$ on (row) probability vectors $\nu \in \mathcal{M}_\Sigma$. Denote by $\mathcal{S}$ the set of all "stationary" probability distributions $A = (a_{ij})$ on $\Omega := \Sigma \times \Sigma$, i.e., $a_{ij} \geq 0$ for all i, j and $\sum_{i,j\in\Sigma} a_{ij} = 1$. Stationarity means here that both marginal distributions of A coincide so that there is a probability vector $\nu^A \in \mathcal{M}_\Sigma$ such that $\nu_j^A = \sum_i a_{ij} = \sum_i a_{ji}$ for all $j \in \Sigma$. (Stationarity in this sense is of course a much weaker property than the stationarity of a stochastic process as usually defined in probability theory.)

Define the entropy of $A \in \mathcal{S}$ as

$$H(A) = -\sum_{i,j\in\Sigma} a_{ij} \log \frac{a_{ij}}{\nu_i^A} = \sum_{i,j\in\Sigma} \nu_i^A \cdot \varphi\left(\frac{a_{ij}}{\nu_i^A}\right) .$$

This is the average amount of information of the events "(i,j) given i". If all ν_i^A are strictly positive, this is clearly well defined. As $0 \leq a_{ij}/\nu_i^A \leq 1$ and as φ is continuous on $[0,1]$, the definition of $H(A)$ extends by continuity to all $A \in \mathcal{S}$. In particular, $H : \mathcal{S} \to \mathbb{R}$ is continuous. We also write $A(\log q_{ij}) = \sum_{i,j} a_{ij} \log q_{ij}$, and as we are going to assume that $q_{ij} > 0$ for all i, j, this is well defined.

1.3.1 Theorem
Suppose that $q_{ij} > 0$ *for all* $i, j \in \Sigma$. *Then*

$$\sup_{A \in \mathcal{S}} \left(H(A) + A(\log q_{ij})\right) = 0\,, \tag{1.9}$$

the supremum is attained for at least one $A \in \mathcal{S}$, *and it is attained for a particular* A *if and only if* $a_{ij} = \nu_i^A q_{ij}$ *for all* $i, j \in \Sigma$. *In this case* $\nu^A Q = \nu^A$.

Proof: As $H : \mathcal{S} \to \mathbb{R}$ is continuous and as all $q_{ij} > 0$, the map $A \mapsto H(A) + A(\log q_{ij})$ is a continuous real-valued map on the compact set $\mathcal{S}$. So it attains its supremum. Let $A \in \mathcal{S}$. Using an elementary version of the Jensen inequality we obtain

$$H(A) + A(\log q_{ij}) = \sum_{\substack{i,j\in\Sigma \\ a_{ij}>0}} a_{ij} \log \frac{\nu_i^A q_{ij}}{a_{ij}} \leq \log \left(\sum_{\substack{i,j\in\Sigma \\ a_{ij}>0}} a_{ij} \frac{\nu_i^A q_{ij}}{a_{ij}} \right) = \log(1) = 0$$

with equality if and only if $\nu_i^A q_{ij} = a_{ij}$ for all $i, j \in \Sigma$. In this case $\sum_i \nu_i^A q_{ij} = \sum_i a_{ij} = \nu_j^A$, i.e., $\nu^A Q = \nu^A$. □

This variational principle characterizes, among all "stationary" probability distributions on $\Sigma \times \Sigma$, just the two-dimensional marginal distributions of the stationary Markov measure for Q. In Section 5.4 we extend this result from distributions on $\Sigma \times \Sigma$ to stationary distributions on $\Sigma^{\mathbb{Z}}$ and characterize stationary Markov measures for the transition matrix Q by a variational principle. There we also show that if all entries of some power of Q are strictly positive, then there is only one such distribution.

1.4 Invariant measures for interval maps

In this section we describe a simple situation where the action of a group or semi-group is interpreted as a dynamical process evolving in discrete time. This is quite different from the interpretation in Section 1.2, where the group was introduced to describe the space homogeneity of a system in equilibrium.

Let $X = [0,1]$ and consider a piecewise differentiable map $T : X \to X$. The n-th iterate of T is denoted by T^n, so $T^n x = T(T^{n-1}x)$. The pair (X, T) is interpreted as a *dynamical system*, i.e., X is a configuration space (the "phase space" of classical mechanics), and T describes the evolution of the system from a configuration x at time n to the configuration Tx at time $n+1$. In more abstract terms one can say that the semigroup $\mathbb{Z}_+$ acts on X via the family $\mathcal{T} = (T^n : n \in \mathbb{Z}_+)$ of transformations.

We are interested in the long time behaviour of the system (X, T). Among the most important dynamical quantities describing this behaviour are the system's upper and lower *Lyapunov exponents*

$$\lambda^+(x) := \limsup_{n\to\infty} \frac{1}{n} \log |(T^n)'x| \,, \quad \lambda^-(x) := \liminf_{n\to\infty} \frac{1}{n} \log |(T^n)'x| \,.$$

The Lyapunov exponents describe the growth of $|(T^n)'x|$ as $n \to \infty$. The question whether the upper and lower exponents coincide can be answered after an application of the chain rule to $(T^n)'$, because limits of the type

$$\lambda(x) := \lim_{n\to\infty} \frac{1}{n} \sum_{k=0}^{n-1} (\log |T'|)(T^k x)$$

are the subject of Birkhoff's ergodic theorem.

1.4.1 Definition *Let $\mathcal{B}$ be the Borel-σ-algebra on X, and let $\mathcal{M}$ be the set of Borel probability measures on X. A measure $\mu \in \mathcal{M}$ is T-invariant if $T\mu = \mu$, i.e., if $\mu(T^{-1}A) = \mu(A)$ for all $A \in \mathcal{B}$. Equivalently: $\int f \circ T \, d\mu = \int f \, d\mu$ for all $f \in C(X)$. $\mathcal{M}(T)$ denotes the set of all T-invariant Borel probability measures on X.*

1.4.2 Theorem (Birkhoff's ergodic theorem)
Suppose $\mu \in \mathcal{M}(T)$ and $f \in L^1_\mu$. Then

$$\bar{f}(x) := \lim_{n\to\infty} \frac{1}{n} \sum_{k=0}^{n-1} f(T^k x)$$

exists for μ-a.e. $x \in X$, $\bar{f}(Tx) = \bar{f}(x)$ for μ-a.e. $x \in X$, and

$$\int_A \bar{f} \, d\mu = \int_A f \, d\mu$$

for each T-invariant measurable set $A \subseteq X$ (i.e., for A satisfying $T^{-1}A = A$).

In Section 2.1 this theorem, which is due to George D. Birkhoff, is proved not only for $\mathbb{Z}_+$-actions but also for $\mathbb{Z}^d_+$-actions, and the underlying space X can of course be much more general than $X = [0, 1]$. Here we note that in the present setting, where the action of T describes the time evolution of a system, the invariance condition $T\mu = \mu$ characterizes temporally stationary states μ of the system. This is in contrast to the situation in Section 1.2 where the G-invariance characterized spatially homogeneous states; an additional stationarity assumption was not necessary there because no temporal dynamics were described. The question, however, whether the "limiting energy values" $u(\omega) = \lim_{n\to\infty} \frac{1}{|\Lambda_n|} \sum_{g\in\Lambda_n} \psi(T^g\omega)$ exist is answered to some extent by the ergodic theorem. Readers who want to know more about the historical origin and the current meaning of the word "ergodic" are referred to Section 2.2.

Birkhoff's theorem has one serious drawback: it makes assertions about the asymptotic behaviour of almost all points with respect to one particular invariant measure. So at least the following two questions must be answered before this theorem can be fruitfully applied:

1. Does T possess an invariant measure at all, i.e., is $\mathcal{M}(T) \neq \emptyset$?
2. Which invariant measure is relevant for the situation under consideration, if there are more than one?

We shall see that the second question is by far the more difficult one, because for most maps $\mathcal{M}(T)$ is an infinite-dimensional subset of $\mathcal{M}$. We answer the first one beforehand:

1.4.3 Theorem (Krylov–Bogolubov)
If $X \neq \emptyset$ is a compact metric space and if $T : X \to X$ is continuous, then $\mathcal{M}(T) \neq \emptyset$.

Proof: Let $\nu \in \mathcal{M}$ be any probability measure on X, e.g., $\nu = \delta_x$ for some $x \in X$. Then

$$\nu_k := \frac{1}{k}\sum_{j=0}^{k-1} T^j\nu \in \mathcal{M}$$

for each $k > 0$. Because of the weak sequential compactness of $\mathcal{M}$ (see A.4.30) there is a subsequence (ν_{k_n}) of (ν_k) that converges weakly to some $\mu \in \mathcal{M}$. This μ is in $\mathcal{M}(T)$, because for each $f \in C(X)$

$$\begin{aligned}
\left|\int f\circ T\,d\mu - \int f\,d\mu\right| &= \lim_{n\to\infty}\left|\int f\circ T\,d\nu_{k_n} - \int f\,d\nu_{k_n}\right| \\
&= \lim_{n\to\infty}\left|\frac{1}{k_n}\sum_{j=0}^{k_n-1}\left(\int f\circ T^{j+1}\,d\nu - \int f\circ T^j\,d\nu\right)\right| \\
&= \lim_{n\to\infty}\frac{1}{k_n}\left|\int f\circ T^{k_n}\,d\nu - \int f\,d\nu\right| \\
&\leq \lim_{n\to\infty}\frac{2}{k_n}\|f\|_\infty = 0\,.
\end{aligned}$$

□

One cannot expect deep insights from a theorem like this, whose proof rests only on the weak compactness of $\mathcal{M}$. In fact, the T-invariant measure μ constructed in the proof of the theorem might, e.g., be the unit mass δ_x in a fixed point $x = Tx$, and for such a measure the assertion of Birkhoff's theorem is trivial. The same is true for the equidistribution on any *periodic orbit* $x, Tx, \ldots, T^{p-1}x, T^p x = x$.

There is one very special case, however, where we can easily prove quite a bit more, namely when $\mathcal{M}(T)$ consists of a single measure μ. In this case we call the pair (X, T) *uniquely ergodic*. See the following exercise for equivalent characterizations of unique ergodicity and Exercise 2.2.9 for a typical example. The reader who wishes to learn more about this type of systems is referred to textbooks on general ergodic theory like [13], [31], [46], [60] and references therein.

1.4.4 Exercise Let T be a continuous map of the compact metric space X. The following assertions are equivalent:

1. (X, T) is uniquely ergodic.
2. For each $f \in C(X)$ the sequence $(\frac{1}{n}\sum_{k=0}^{n-1} f(T^k x))_{n>0}$ converges pointwise to a constant.
3. For each $f \in C(X)$ the sequence $(\frac{1}{n}\sum_{k=0}^{n-1} f \circ T^k)_{n>0}$ converges uniformly to a constant. ◇

The much more subtle problem of choosing a "relevant" invariant measure for T leads again to a variational principle and thus to the main theme of this book. Before discussing this problem by means of a simple example I want to present some heuristic thoughts about what "relevant" could mean. Dynamical systems (X, T) often arise from the need to model some process from the "real world". A researcher tries to capture some essential features in a mathematical model, hoping that properties of the model will reflect and explain properties of the real situation. If a skilled researcher has chosen a model (X, T) consisting of a (piecewise) smooth manifold and a (piecewise) smooth map, the underlying Euclidean (or, more generally, Riemannian) structure of the model is likely to reflect an aspect of the real situation that is important to him. Therefore it seems reasonable to expect that a "relevant" invariant measure is tied in some way to this metric structure. The most straightforward (but by no means the only) way to postulate such a connection is to require that the measure is absolutely continuous with respect to Lebesgue measure (or to the Riemannian volume on X). The next example illustrates how such an invariant measure can be distinguished by a variational principle.

1.4.5 Example Let $a, b > 1$, $\frac{1}{a} + \frac{1}{b} = 1$ and $c = \frac{1}{a}$. We study the *full skew tent map*

$$T(x) = \begin{cases} ax & \text{if } x \leq c, \\ 1 - b(x - c) & \text{if } x \geq c. \end{cases}$$

In order to describe its dynamical properties we introduce a kind of *symbolic dynamics* for it: denoting by $X_0 \subset [0,1]$ the set of all points x for which $T^n x \neq c$ for all $n \geq 0$, we define a map

$$\Phi : X_0 \to \Omega := \{0,1\}^{\mathbb{Z}_+}, \quad (\Phi x)_n = \begin{cases} 0 & \text{if } T^n x < c, \\ 1 & \text{if } T^n x > c, \end{cases}$$

and observe:

▷ Φ is injective. To see this, consider two points $x < y$ with $(\Phi(x))_j = (\Phi(y))_j$ for $j = 0, \ldots, n-1$. This means that the restriction of the iterate T^n to the interval (x,y) is monotone so that

$$1 \geq |T^n y - T^n x| = |y - x| \cdot |(T^n)'_{|(x,y)}| \geq |y - x| \cdot \min\{a^n, b^n\}.$$

Hence $|x - y| \leq \max\{c^n, (1-c)^n\}$, and if $\Phi(x) = \Phi(y)$, then $x = y$. At the same time this proves the continuity of $\Phi^{-1} : \Phi(X_0) \to X_0$ when $\Phi(X_0) \subset \Omega$ is endowed with the metric $d(\omega, \omega')$ introduced at the end of Section 1.2, because $d(\omega, \omega') \leq 2^{-n}$ is equivalent to $(\omega)_j = (\omega')_j$ for $j = 0, \ldots, n-1$.

▷ Φ is "nearly surjective" in the sense that the set $\Omega \setminus \Phi(X_0)$ consists of the countably many points $\omega = (\omega_0\omega_1 \ldots \omega_k 10^\infty) \in \Omega$. This is reminiscent of the nonuniqueness of p-ary number representations as, e.g., in binary representation $0.a_1 \ldots a_k 0111 \ldots = 0.a_1 \ldots a_k 1000 \ldots$ holds.

Let $S : \Omega \to \Omega$, $(S\omega)_i = \omega_{i+1}$ denote the *shift transformation*. Then $S \circ \Phi = \Phi \circ T$. As Φ is a bijection between X_0 and $\Phi(X_0)$ (and that means between X and Ω except for countably many points), we say that S and T are *conjugate*.

Let ν be some stationary, non-atomic probability measure on Ω. (Stationarity of ν means $S\nu = \nu$.) We want to define a probability measure $\Phi^{-1}\nu$ on $[0,1]$ by $\Phi^{-1}\nu(A) = \nu(\Phi(A \cap X_0))$. As Φ^{-1} is continuous and hence measurable on its domain, this is a well defined measure, and as $\Omega \setminus \Phi(X_0)$ is countable, $\Phi^{-1}\nu([0,1]) = \nu(\Phi(X_0)) = 1$ so that $\Phi^{-1}\nu$ is indeed a probability measure. It is T-invariant, because $\Phi^{-1} \circ S = T \circ \Phi^{-1}$ on $\Phi(X_0)$ so that $T\Phi^{-1}\nu = \Phi^{-1}S\nu = \Phi^{-1}\nu$.

At this introductory level we restrict ourselves to very particular stationary measures on Ω, namely to Bernoulli measures P_q with parameter q. These are the measures under which the ω_i, $i \in \mathbb{Z}_+$, are independent and distributed with $P_q(\omega_i = 1) = q$, $P_q(\omega_i = 0) = 1 - q$. Let $\mu_q := \Phi^{-1}P_q$ and define $g : \Omega \to \mathbb{R}$ by

$$g(\omega) := g_0(\omega_0) := \begin{cases} \log a & \text{if } \omega_0 = 0, \\ \log b & \text{if } \omega_0 = 1. \end{cases}$$

Then $\log|T'| = g \circ \Phi$, and it follows from Birkhoff's ergodic theorem that the Lyapunov exponent

$$\begin{aligned} \lambda(x) &= \lim_{n\to\infty} \frac{1}{n} \sum_{k=0}^{n-1} \log|T'(T^k x)| = \lim_{n\to\infty} \frac{1}{n} \sum_{k=0}^{n-1} g(\Phi(T^k x)) \\ &= \overline{g \circ \Phi}(x) = \overline{\log|T'|}(x) \end{aligned}$$

for μ_q-a.e. x. In order to identify $\overline{\log|T'|}$ as $\mu_q(\log|T'|)$ we use the product structure of the measures P_q: as $g(\Phi(T^k x)) = g(S^k(\Phi x)) = g_0((\Phi x)_k)$ and as $(\Phi x)_0, (\Phi x)_1, (\Phi x)_2, \dots$ are independent and identically distributed random variables under $\mu_q = \Phi^{-1}P_q$, it follows from the strong law of large numbers that

$$\lambda(x) = \lim_{n\to\infty} \frac{1}{n}\sum_{k=0}^{n-1} g_0((\Phi x)_k) = (1-q)\log a + q\log b = P_q(g) = \mu_q(\log|T'|)$$

for μ_q-a.e. $x \in X$. So $\overline{\log|T'|} = \mu_q(\log|T'|)$ is constant μ_q-a.e. (Later we say that this is a consequence of the *ergodicity* of μ_q.)

Let $H(q) = -(q\log q + (1-q)\log(1-q))$. This is an expression for the *entropy* of a Bernoulli process that will be justified in Chapter 3, see Example 3.2.20. As in the proof of Theorem 1.1.3, the elementary version of Jensen's inequality and the fact that $\frac{1}{a} + \frac{1}{b} = 1$ imply

$$\begin{aligned} H(q) - \mu_q(\log|T'|) &= (1-q)\log\frac{1}{(1-q)a} + q\log\frac{1}{qb} \\ &\le \log\left(\frac{1-q}{(1-q)a} + \frac{q}{qb}\right) = \log(1) = 0 \end{aligned}$$

with equality if and only if $(1-q)a = qb$, i.e., if $q = \frac{1}{b} = 1-c$ and $1-q = \frac{1}{a} = c$.

We turn to the second aspect under which $q = 1-c$ is a distinguished parameter: among the measures μ_q, which are all mutually singular, μ_{1-c} is the only one that is absolutely continuous with respect to Lebesgue measure m on $[0,1]$. (In fact, it is a special feature of this particular example that $\mu_{1-c} = m$.) The mutual singularity of the μ_q is an immediate consequence of the mutual singularity of the Bernoulli measures P_q. To see that $\mu_{1-c} = m$, consider *cylinder sets* $[\sigma_0 \dots \sigma_{n-1}] := \{\omega \in \Omega : \omega_i = \sigma_i \ (i = 0, \dots, n-1)\}$. For each choice of $\sigma_0, \dots, \sigma_{n-1} \in \{0,1\}$ the set $\Phi^{-1}[\sigma_0 \dots \sigma_{n-1}]$ is an interval of length $P_{1-c}[\sigma_0 \dots \sigma_{n-1}]$, because it is mapped onto $[0,1]$ by a linear branch of T^n with slope $a^k b^{n-k} = (P_{1-c}[\sigma_0 \dots \sigma_{n-1}])^{-1}$ where k is the number of zeros among $\sigma_0, \dots, \sigma_{n-1}$.

As a glimpse of more general results in Section 6.3, we summarize:

▷ The measure μ_{1-c}, which is absolutely continuous with respect to *Lebesgue measure*, maximizes the expression $H(q) + \mu_q(-\log|T'|)$, where $|T'|$ is the derivative of T with respect to *Lebesgue measure* (in the sense that $m(TA) = \int_A |T'|\,dm$ for each measurable set A on which T is injective). In the terminology of Section 1.2, the measure μ_{1-c} is an equilibrium state for the function $\psi = -\log|T'|$.

▷ $\lambda(x) = H(1-c)$ for Lebesgue-a.e. $x \in X$.

▷ If one is interested in the Lebesgue null set of points with "untypical" Lyapunov exponents $\lambda(x) \neq H(c)$, the other measures μ_q also play a role. In fact, they are equilibrium states for $\psi = -\beta \log |T'|$ with $\beta \neq 1$, and their properties are closely related to the considerations on large deviations in Section 1.2, see also Exercise 1.4.7. More on large deviations for interval maps and other dynamical systems can be found in Proposition 6.1.11. ◇

1.4.6 Remark In the Appendix, Section A.5, a mathematically more precise version of the above map Φ is introduced. By adding countably many points to $[0,1]$ the interval is enlarged to a compact phase space X, and Φ is extended to all of X in such a way that $\Phi : X \to \Omega$ becomes a homeomorphism. In this way one can also use the Krylov–Bogolubov theorem to show the existence of at least one invariant measure for most piecewise continuous maps. ◇

1.4.7 Exercise In the situation of Example 1.4.5 let $q_0 \in (0,1)$ and determine $\beta_0 \in \mathbb{R}$ such that $H(q) - \beta_0 \mu_q(\log |T'|)$ is maximized for $q = q_0$. ◇

1.4.8 Exercise Let $a, b > 1$, $\frac{1}{a} + \frac{1}{b} = 1$ and $c = \frac{1}{a}$. Define the *baker's map* T on $[0,1]^2$ by

$$T(x,y) = \begin{cases} (ax, \frac{y}{a}) & \text{if } x \leq c, \\ (b(x-c), c + \frac{y}{b}) & \text{if } x > c. \end{cases}$$

Along the lines of Example 1.4.5 show that the baker's map is conjugate to the shift transformation on the two-sided sequence space $\Omega = \{0,1\}^{\mathbb{Z}}$ by a conjugating map Φ that transforms the Lebesgue measure on $[0,1]^2$ to the Bernoulli measure P_c on Ω. (Here "conjugating" is meant in the sense that there are measurable subsets $X_0 \subseteq [0,1]$ and $\Omega_0 \subseteq \Omega$ of full Lebesgue and P_c-measure respectively and such that Φ is a bimeasurable bijection from X_0 onto Ω_0.)

Chapter 2

Some basic ergodic theory

Ergodic theory deals with groups or semigroups of measure preserving transformations T^g on a probability space $(X, \mathcal{B}, \mu)$. In later sections X will be a compact metric space endowed with its Borel σ-algebra $\mathcal{B}$, but most of the theory presented in this chapter is independent of this topological background.

We start with a section on Birkhoff's ergodic theorem, which generalizes the classical formulation from Theorem 1.4.2 to arbitrary measure preserving $\mathbb{Z}^d_+$-actions. In order to characterize those systems for which the limit in the ergodic theorem is constant, we introduce the concept of ergodicity in Section 2.2 and discuss the related notions of weak mixing and mixing. In Section 2.3 we show how an arbitrary measure preserving system can be decomposed into ergodic components. We finish this chapter with two brief sections on return times and return maps and on factors and extensions. Both may be skipped on a first reading, because we make no explicit use of them except in Chapter 6.

There are several textbooks on general ergodic theory which the reader might wish to consult for further information, for example [60], [13], [46], [37].

2.1 Birkhoff's ergodic theorem

The basic objects of ergodic theory are measure preserving dynamical systems, a notion that embraces as special cases lattice systems discussed at the end of Section 1.2 and the interval maps from Section 1.4. In this section we state and prove Birkhoff's ergodic theorem for $\mathbb{Z}^d_+$-actions that applies to such systems.

2.1.1 Definition *A measure preserving dynamical system (m.p.d.s.) is a quadruple $(X, \mathcal{B}, \mu, \mathcal{T})$ consisting of*

a) a probability space $(X, \mathcal{B}, \mu)$ and

b) a $\mathcal{B}$-measurable action $\mathcal{T} = (T^g : g \in G)$ of the semigroup $G = \mathbb{Z}^d_+$ or of the group $G = \mathbb{Z}^d$ on X which leaves the measure μ invariant (symbolically: $\mathcal{T}\mu = \mu$). This means:

1) $T^g : X \to X$ is $\mathcal{B}$-measurable and $T^g\mu = \mu$ for all $g \in G$,
2) $T^0 = Id_X$ and $T^{g+g'} = T^g \circ T^{g'}$ for all $g, g' \in G$.

2.1.2 Remarks

a) Observe that $\mathcal{T}\mu = \mu$ implies

$$\int_{T^{-g}A} f \circ T^g \, d\mu = \int_A f \, d\mu, \text{ in particular } \|f \circ T^g\|_1 = \|f\|_1$$

for all $A \in \mathcal{B}$, $f \in L^1_\mu$ and $g \in G$.

b) If $G = \mathbb{Z}^d$, then all $T^g \in \mathcal{T}$ are invertible, because $-g \in G$ for all $g \in G$ and $T^g \circ T^{-g} = T^0 = Id_X$.

c) Let $e_1, \ldots, e_d$ be the canonical basis for the lattice $\mathbb{Z}^d_+$ and consider $g = \sum_{i=1}^d g_i e_i \in G$. Then $T^g = (T^{e_1})^{g_1} \circ \cdots \circ (T^{e_d})^{g_d}$ where $(T^{e_i})^{g_i}$ denotes the g_i-fold iterate of T^{e_i}. Hence, if $d = 1$, then T^g is the g-fold iterate of the map $T = T^1$. In this case we also write $(X, \mathcal{B}, \mu, T)$ instead of $(X, \mathcal{B}, \mu, \mathcal{T})$. If $d > 1$, the transformations T^{e_i} do not necessarily commute.

d) Recent contributions to the ergodic theory of $\mathbb{Z}^d$-actions are collected in [48]. Actions of more general groups are studied in [42], [54]. ◇

2.1.3 Definition *A measurable function $f : X \to \mathbb{R}$ is $\mathcal{T}$-invariant* mod μ *if $f \circ T^g = f$ μ-a.s. for each $g \in G$. Accordingly a set $A \in \mathcal{B}$ is $\mathcal{T}$-invariant* mod μ *if $\mu(T^{-g}A \triangle A) = 0$ for all $g \in G$. We write $\mathcal{I}_\mu(\mathcal{T}) := \{A \in \mathcal{B} : \mu(T^{-g}A \triangle A) = 0 \; \forall g \in G\}$. If $\mathcal{T}$ is generated by a single transformation T, we write also $\mathcal{I}_\mu(T)$ instead of $\mathcal{I}_\mu(\mathcal{T})$.*

2.1.4 Remark Let $\mathcal{I}(\mathcal{T}) := \{B \in \mathcal{B} : T^{-g}B = B \; \forall g \in G\}$. Then $\mathcal{I}(\mathcal{T}) = \mathcal{I}_\mu(\mathcal{T})$ mod μ, because for each $B \in \mathcal{I}_\mu(\mathcal{T})$ the set $B' := \bigcup_{g \in G} \bigcap_{h \in G} T^{-(g+h)}B$ belongs to $\mathcal{I}(\mathcal{T})$ and $\mu(B \triangle B') = 0$.

It is easy to show that $\mathcal{I}(\mathcal{T})$ and $\mathcal{I}_\mu(\mathcal{T})$ are σ-algebras and that a $\mathcal{B}$-measurable function f is $\mathcal{T}$-invariant ($\mathcal{T}$-invariant mod μ) if and only if it is $\mathcal{I}(\mathcal{T})$-measurable ($\mathcal{I}_\mu(\mathcal{T})$-measurable). ◇

For $n \in \mathbb{Z}_+$ let

$$\Lambda_n := \{g = \sum_i g_i e_i \in G : |g_i| < n \; \forall i = 1, \ldots, d\} \quad \text{and} \quad \lambda_n := |\Lambda_n| .$$

One of the corner-stones of ergodic theory is

2.1.5 Theorem (Birkhoff's ergodic theorem)
Let $(X, \mathcal{B}, \mu, \mathcal{T})$ be a m.p.d.s. For each $f \in L^1_\mu$ the limit

$$\bar{f}(x) := \lim_{n\to\infty} \frac{1}{\lambda_n} \sum_{g\in\Lambda_n} f(T^g x) \tag{2.1}$$

exists μ-a.s. and in L^1_μ. The function $\bar{f}$ is $\mathcal{T}$-invariant mod μ *and for each set $A \in \mathcal{I}_\mu(\mathcal{T})$*

$$\int_A \bar{f}\, d\mu = \int_A f\, d\mu\,. \tag{2.2}$$

2.1.6 Remarks
a) The limit $\bar{f}$ is $\mathcal{I}_\mu(\mathcal{T})$-measurable, and in view of identity (2.2) it is a version of $E_\mu[f \mid \mathcal{I}_\mu(\mathcal{T})]$, the conditional expectation of f under μ given $\mathcal{I}_\mu(\mathcal{T})$, see A.4.32.

b) It suffices to prove the theorem for $G = \mathbb{Z}^d_+$, because the case $G = \mathbb{Z}^d$ can be reduced to $G = \mathbb{Z}^d_+$ in the following way:

Let $\mathcal{T} = (T^g : g \in \mathbb{Z}^d)$ be a $\mathbb{Z}^d$-action. For each $\sigma \in \{+1, -1\}^d$ denote by $\mathcal{T}_\sigma = (T^g_\sigma : g \in \mathbb{Z}^d_+)$ the $\mathbb{Z}^d_+$-action $T^g_\sigma := T^{(\sigma_1 g_1, \dots, \sigma_d g_d)}$. Since a measurable set is T^g-invariant mod μ if and only if it is T^{-g}-invariant mod μ, the σ-algebras $\mathcal{I}_\mu(\mathcal{T}_\sigma)$ are all identical and coincide with $\mathcal{I}_\mu(\mathcal{T})$. Now apply the ergodic theorem for the action $\mathcal{T}_\sigma$ to a function $f \in L^1_\mu$ and denote the resulting limit by $\bar{f}_\sigma$. By the previous remark, all these $\bar{f}_\sigma$ are versions of the conditional expectation $E_\mu[f \mid \mathcal{I}_\mu(\mathcal{T})]$.

Denote now by Λ_n the n-box in $\mathbb{Z}^d$ and by Λ^+_n the one in $\mathbb{Z}^d_+$. If it were true that $\sum_{g\in\Lambda_n} f \circ T^g = \sum_\sigma \sum_{g\in\Lambda^+_n} f \circ T^g_\sigma$, it would follow immediately that $\bar{f} = 2^{-d} \sum_\sigma \bar{f}_\sigma = E_\mu[f \mid \mathcal{I}_\mu(\mathcal{T})]$ μ-as. However, this decomposition of the summation over Λ_n counts elements g with exactly one $g_i = 0$ twice and, more generally, those with exactly k indices i for which $g_i = 0$ are counted 2^k times, see Figure 2.1. Therefore the sum $\sum_\sigma \sum_{g\in\Lambda^+_n} f \circ T^g_\sigma$ must be modified by averages over sets of elements $g \in \Lambda^+_n$ for which certain g_i vanish. But this means that these averages obey the ergodic theorem for a $\mathbb{Z}^{d'}_+$-action with $d' < d$. In particular, the corresponding norming constants are $n^{-d'}$ so that these contributions vanish asymptotically if they are normed by $\lambda_n^{-1} = n^{-d}$.

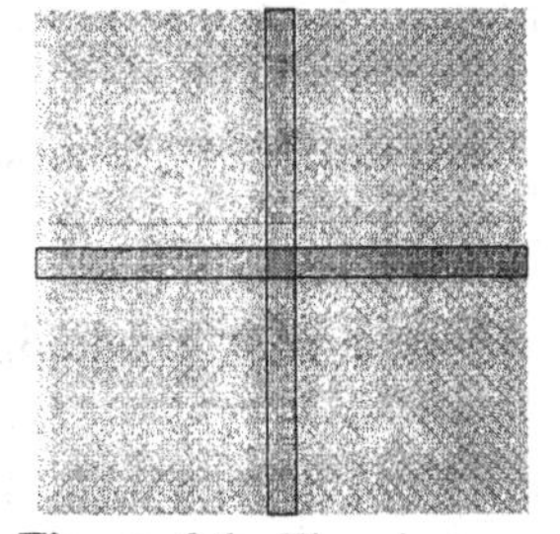

Figure 2.1: The decomposition of Λ_n

c) In the one-dimensional case the ergodic theorem takes the form

$$\bar{f}(x) = \lim_{n\to\infty} \frac{1}{n} \sum_{k=0}^{n-1} f(T^k x) \quad \text{for } \mu\text{-a.e. } x.$$

By the previous remark one has for $G = \mathbb{Z}$

$$\bar{f}(x) = \lim_{n\to\infty} \frac{1}{n} \sum_{k=0}^{n-1} f(T^k x) = \lim_{n\to\infty} \frac{1}{n} \sum_{k=0}^{n-1} f(T^{-k} x) \quad \mu\text{-a.s.}$$

d) Ergodic averages along more general families of "parameter intervals" than Λ_n are also of interest. Theorem 6.2.8 in [34] shows that these averages converge to essentially the same $\bar{f}$ for a broad class of d-dimensional index sets.

◇

Proof of the ergodic theorem: This proof comes in several steps. Some of them are independent of the dimension, others are much simpler for $d = 1$ than for $d > 1$. In order not to overload the proof with technicalities we therefore perform some of the steps only for $d = 1$ first and provide the necessary modifications for $d > 1$ later.

For $d = 1$ we are following closely the proof in [4, Theorem 2.2]. Steps 1, 2, and 4 of our exposition are written in such a way that we can refer to them later in our proof for the case $d > 1$, which is close to the proof of [34, Theorem 6.2.8]. Also, for $d > 1$, we use the ergodic theorem for $\mathbb{Z}_+^{d-1}$-actions in Step 1. This introduces some inductive structure to the proof.

We use the following notation: for measurable $f : X \to \mathbb{R}$ let

$$S_n f := \sum_{g\in\Lambda_n} f \circ T^g , \quad A_n f := \frac{1}{\lambda_n} S_n f ,$$

$$A^- f := \liminf_{n\to\infty} A_n f , \quad A^+ f := \limsup_{n\to\infty} A_n f .$$

Step 1: As $f = f^+ - f^-$, it suffices to prove the theorem for $0 \le f \in L^1_\mu$. For $i = 1, \dots, d$ we have

$$\begin{aligned} A_n f \circ T^{e_i} &= A_n f - \frac{1}{n^d} \sum_{g\in\Lambda_n, g_i=0} f \circ T^g + \frac{1}{n^d} \sum_{g\in(\Lambda_n+e_i)\setminus\Lambda_n} f \circ T^g \\ &\ge A_n f - \frac{1}{n} \left(\frac{1}{n^{d-1}} \sum_{g\in\Lambda_n, g_i=0} f \circ T^g \right) . \end{aligned}$$

If $d = 1$ (and hence $i = 1$), the term in brackets is identically equal to f. If $d > 1$ we apply the ergodic theorem for $\mathbb{Z}_+^{d-1}$-actions to this term and conclude that it converges almost surely to some finite value. Therefore, in any case

$$A^+ f \circ T^{e_i} = \limsup_{n\to\infty} A_n f \circ T^{e_i} \ge \limsup_{n\to\infty} A_n f = A^+ f$$

μ-a.s. As at the same time $S_n f \circ T^{e_i} \le S_{n+1} f$ and hence

$$A^+ f \circ T^{e_i} = \limsup_{n\to\infty} \frac{1}{\lambda_n} S_n f \circ T^{e_i} \le \limsup_{n\to\infty} \frac{(n+1)^d}{n^d} \frac{1}{\lambda_{n+1}} S_{n+1} f = A^+ f ,$$

it follows that

$$A^+ f \circ T^g = A^+ f \quad \mu\text{-a.s.} \tag{2.3}$$

for all $g \in G$.

Step 2: The idea of the proof is to show rather directly that $A^- f \geq A^+ f$ μ-a.s. We will find, however, that such a naive approach does not yield a pathwise estimate. Instead we are going to prove that

$$\int A^+ f \, d\mu \leq \gamma_d \int f \, d\mu$$

where $\gamma_1 = 1$ and $\gamma_d = 2^d$ for $d > 1$. For the case $d = 1$ it is then an easy task to deduce from this in Step 3 that $\int A^+ f \, d\mu \leq \int A^- f \, d\mu$ and hence $A^+ f = A^- f$ μ-a.s.

The basic idea is to decompose the sum $S_n f(x)$ into sums over small cubes Λ_k (uniformly bounded in size) in such a way that the averages of f over these subcubes are nearly equal to $A^+ f(x)$.

For two real-valued functions g, h on X let $g \wedge h := \min\{g, h\}$ be the pointwise minimum. Let $r > 1$, $0 < \epsilon < 1$, and define $H = H_{r,\epsilon} := (A^+ f \wedge r)(1 - \epsilon)$. Then $0 \leq H < r$, and as $(A^+ f) \circ T^g = A^+ f$ by (2.3), also $H \circ T^g = H$ for all $g \in G$. In order to perform the above mentioned decomposition let

$$\tau : X \to \mathbb{N}, \quad x \mapsto \min\{n \geq 1 : S_n f(x) \geq \lambda_n \cdot H(x)\} .$$

Observe that $\tau(x) < \infty$ for all $x \in X$:

if $A^+ f(x) = 0$, then $S_1 f(x) = f(x) \geq 0 = H(x)$ so that $\tau(x) = 1$,
if $0 < A^+ f(x) < \infty$, then $A^+ f(x) > H(x)$ so that $\tau(x) < \infty$, and
if $A^+ f(x) = \infty$, then $A^+ f(x) > r(1 - \epsilon) = H(x)$ so that $\tau(x) < \infty$.

The definition of τ suggests that the subblocks of Λ_n we are looking for are of the form $\Lambda_{\tau(T^g x)} + g$ for suitable $g \in \Lambda_n$. If τ is uniformly bounded we can actually work with such blocks. Otherwise we replace f by a function $\tilde{f}$ close to f such that the associated $\tilde{\tau}$ is bounded. More precisely, for $M > 0$ consider $\tilde{f} := f + H \cdot 1_{\{\tau > M\}}$ and $\tilde{\tau}(x) := \min\{n \geq 1 : S_n \tilde{f}(x) \geq \lambda_n \cdot H(x)\}$. Observe that

if $\tau(x) > M$, then $S_1 \tilde{f}(x) = \tilde{f}(x) \geq H(x)$, i.e., $\tilde{\tau}(x) = 1 \leq M$, and
if $\tau(x) \leq M$, then $S_{\tau(x)} \tilde{f}(x) \geq S_{\tau(x)} f(x) \geq \lambda_{\tau(x)} H(x)$, i.e., $\tilde{\tau}(x) \leq \tau(x) \leq M$.

Hence $\tilde{\tau} \leq M$. Suppose now that for each $x \in X$ there is some set $\Lambda'_n(x) \subseteq \Lambda_n$ with the property that $(\Lambda_{\tilde{\tau}(T^g x)} + g : g \in \Lambda'_n(x))$ is a family of pairwise disjoint "cubes" contained in Λ_n such that

$$\sum_{g \in \Lambda'_n(x)} |\Lambda_{\tilde{\tau}(T^g x)}| \geq \gamma_d^{-1} (n - M)^d . \tag{2.4}$$

This condition ensures that the union of these cubes covers approximately a fraction γ_d^{-1} of Λ_n at least, see Figure 2.2. For general d this family is provided by Lemma 2.1.7 below; in the case $d = 1$ the set $\Lambda'_n(x)$ is simply the set of all integers $\tau_k(x) \in \Lambda_{n-M}$ defined recursively by

$$\tau_0 = 0 \quad \text{and} \quad \tau_{k+1} = \tau_k + \tilde{\tau} \circ T^{\tau_k} \text{ for } k \geq 0 .$$

In view of the definition of $\tilde{\tau}$ and the $\mathcal{T}$-invariance of H we can estimate

$$\begin{aligned} S_n\tilde{f}(x) &\geq \sum_{h\in\Lambda'_n(x)} \sum_{g\in\Lambda_{\tilde{\tau}(T^h x)}+h} (\tilde{f}\circ T^g)(x) = \sum_{h\in\Lambda'_n(x)} \sum_{g\in\Lambda_{\tilde{\tau}(T^h x)}} (\tilde{f}\circ T^g)(T^h x) \\ &= \sum_{h\in\Lambda'_n(x)} (S_{\tilde{\tau}(T^h x)}\tilde{f})(T^h x) \geq \sum_{h\in\Lambda'_n(x)} \lambda_{\tilde{\tau}(T^h x)} \cdot H(T^h x) \\ &= \sum_{h\in\Lambda'_n(x)} |\Lambda_{\tilde{\tau}(T^h x)}| \cdot H(x) \geq \gamma_d^{-1}(n-M)^d \cdot H(x) . \end{aligned}$$

Hence, observing that $H \leq r$,

$$S_n f = S_n(\tilde{f} - H\,1_{\{\tau>M\}}) \geq \gamma_d^{-1}(n-M)^d \cdot H - r\,S_n 1_{\{\tau>M\}} . \tag{2.5}$$

Dividing this inequality by λ_n and passing to the limit $n \to \infty$ this yields at once that $A^- f \geq \gamma_d^{-1} H - rA^+ 1_{\{\tau>M\}} = \gamma_d^{-1}(1-\epsilon)(A^+ f \wedge r) - rA^+ 1_{\{\tau>M\}}$. The hope is now that $A^+ 1_{\{\tau>M\}}$ is small because $\mu\{\tau > M\}$ is small if M is large enough. But this term is of the same type as the term $A^+ f$ that we want to estimate. Furthermore, we have made no use of the $\mathcal{T}$-invariance of μ so far, and it is exactly at this point where we have to use it. Instead of estimating pathwise averages $A_n 1_{\{\tau>M\}}(x)$, we integrate inequality (2.5) and observe that $\mathcal{T}\mu = \mu$:

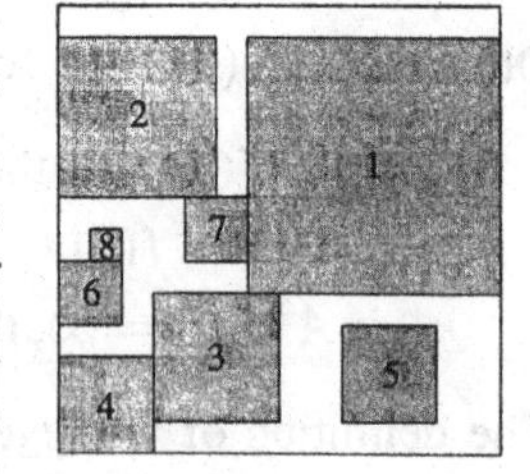

Figure 2.2: An example for the cubes $\Lambda_{\tau(T^g x)} + g$

$$\int f\,d\mu = \frac{1}{n^d} \int S_n f\,d\mu \geq \gamma_d^{-1} \left(\frac{n-M}{n}\right)^d \int H\,d\mu - r\,\mu\{\tau > M\} .$$

In the limit $n \to \infty$ this yields

$$\int f\,d\mu + r\,\mu\{\tau > M\} \geq \gamma_d^{-1} \int H\,d\mu .$$

As $\lim_{M\to\infty} \mu\{\tau > M\} = 0$ and as $H = (1-\epsilon)(A^+ f \wedge r)$, this yields

$$\gamma_d \int f\,d\mu \geq (1-\epsilon) \int (A^+ f \wedge r)\,d\mu .$$

Passing with $\epsilon \to 0$ and then with $r \to \infty$ it follows that

$$\int A^+ f \, d\mu \le \gamma_d \int f \, d\mu \,. \tag{2.6}$$

(The last limit uses the monotone convergence theorem, see A.4.19.)

Step 3 (only for $d = 1$): Here we derive the almost sure convergence in (2.1) from estimate (2.6). Our first observation is that for bounded f ($0 \le f \le M$) we have $A^- f = -A^+(-f) = M - A^+(M - f)$ so that

$$\int A^- f \, d\mu = M - \int A^+(M - f) \, d\mu \ge M - \int (M - f) \, d\mu = \int f \, d\mu$$

by (2.6) as $\gamma_1 = 1$. By a simple truncation argument this inequality extends to unbounded $f \ge 0$: Let $M > 0$. Then

$$\int A^- f \, d\mu \ge \int A^-(f \wedge M) \, d\mu \ge \int f \wedge M \, d\mu \nearrow \int f \, d\mu$$

as $M \to \infty$ by the monotone convergence theorem (see A.4.19), so that for any measurable $f \ge 0$

$$\int A^- f \, d\mu \ge \int f \, d\mu \ge \int A^+ f \, d\mu \,. \tag{2.7}$$

As $A^- f \le A^+ f$ it follows that $\bar{f} = A^- f = A^+ f$ μ-a.s.

Step 4: We prove the remaining assertions of the theorem. For the L^1_μ-convergence of $A_n f$ to $\bar{f}$ it suffices to show that the sequence $(A_n f)_n$ is uniformly integrable (see A.4.21): as $A_n f - r = A_n(f - r) \le A_n((f - r)^+)$ where $A_n((f - r)^+) \ge 0$, we have also $(A_n f - r)^+ \le A_n((f - r)^+)$ so that

$$\int (A_n f - r)^+ \, d\mu \le \int A_n((f - r)^+) \, d\mu = \int (f - r)^+ d\,\mu \,,$$

and this tends to zero as $r \to \infty$.

The $\mathcal{T}$-invariance mod μ of $\bar{f}$ follows from (2.3), and (2.2) is now a direct consequence of the $\mathcal{T}$-invariance of μ: if $A \in \mathcal{I}_\mu(\mathcal{T})$, then

$$\int_A \bar{f} \, d\mu = \lim_{n\to\infty} \frac{1}{\lambda_n} \int_A S_n f \, d\mu = \lim_{n\to\infty} \frac{1}{\lambda_n} \sum_{g \in \Lambda_n} \int_{T^{-g}A} f \circ T^g \, d\mu = \int_A f \, d\mu \,.$$

This finishes the proof of the ergodic theorem for the case $d = 1$. Reviewing the four steps we see that the following modifications and additions are necessary for $d > 1$:

1. a different choice of the cubes in inequality (2.4), and
2. a new proof of $A^+f = A^-f$ μ-a.s. in Step 3.

We start with the first point. In the case $d > 1$, the cube Λ_n can no longer be paved without gaps using cubes $\Lambda_{\tau(T^h x)}+h$ as in one dimension. Still the following lemma provides a sufficient (though fragmentary) paving:

2.1.7 Lemma *Given positive integers $n > M$ and a map $\ell : \Lambda_n \to \{1, \dots, M\}$, there is a set $\Lambda'_n \subseteq \Lambda_n$ such that $(\Lambda_{\ell(g)}+g : g \in \Lambda'_n)$ is a family of pairwise disjoint "cubes" contained in Λ_n and*

$$\sum_{g\in\Lambda'_n} |\Lambda_{\ell(g)}| \geq 2^{-d}(n-M)^d . \tag{2.8}$$

Proof: Let $\mathcal{D}_M$ be a maximal collection of disjoint sets of the form $\Lambda_M + g$ with $\ell(g) = M$ and $g \in \Lambda_{n-M}$. When $\mathcal{D}_M, \dots, \mathcal{D}_i$ have been constructed for some $i > 1$, let $\mathcal{D}_{i-1}$ be a maximal collection of sets of the form $\Lambda_{i-1} + g$ with $\ell(g) = i-1$ and $g \in \Lambda_{n-M}$, for which all sets in $\mathcal{D}_{i-1} \cup \dots \cup \mathcal{D}_M$ are disjoint. Let

$$\Lambda'_n := \{g \in \Lambda_{n-M} : \Lambda_{\ell(g)} + g \in \mathcal{D}_1 \cup \dots \cup \mathcal{D}_M\} .$$

Observe that the sets $D_g := \Lambda_{\ell(g)} + g$, $g \in \Lambda'_n$, are pairwise disjoint subsets of Λ_n.

Take any $h \in \Lambda_{n-M}$. By the maximality of $\mathcal{D}_{\ell(h)}$ there exist some $j \geq \ell(h)$ and $\Lambda_j + g \in \mathcal{D}_j$ with $(\Lambda_{\ell(g)} + g) \cap (\Lambda_{\ell(h)} + h) \neq \emptyset$. Then $\ell(g) = j \geq \ell(h)$ and $\Lambda_{\ell(g)} \supseteq \Lambda_{\ell(h)}$. Hence $h \in \tilde{D}_g := g+\Lambda_{\ell(g)}-\Lambda_{\ell(g)}$, and therefore $\Lambda_{n-M} \subseteq \bigcup_{g\in\Lambda'_n} \tilde{D}_g$. Now (2.8) follows from the inequality $|\tilde{D}_g| \leq 2^d \cdot |D_g|$. □

It remains to replace Step 3 in the proof of the ergodic theorem by a more elaborate argument. For the following lemma we need the concept of an orthogonal complement in the Hilbert space L^2_μ: For a linear subspace N of L^2_μ let $N^\perp := \{h \in L^2_\mu : \langle f, h\rangle = 0\, \forall f \in N\}$. Then $N^\perp$ is a closed linear subspace and $\mathrm{clos}(N) \oplus N^\perp = L^2_\mu$.

2.1.8 Lemma *Let $\mathcal{T} = \{T^g : g \in G\}$, $F := \{f \in L^2_\mu : f \circ T = f\; \forall T \in \mathcal{T}\}$ and $N := \{h - h\circ T : h \in L^2_\mu, T \in \mathcal{T}\}$. Then $F = N^\perp$.*

Proof: To show $F \subseteq N^\perp$, let $f \in F$, $h \in L^2_\mu$ and $T \in \mathcal{T}$. Then

$$\langle f, h - h\circ T\rangle = \langle f, h\rangle - \langle f\circ T, h\circ T\rangle = \langle f, h\rangle - \langle f, h\rangle = 0 ,$$

i.e., $f \in N^\perp$.

For the reverse inclusion let $f \in N^\perp$. Then $\langle f, f - f\circ T\rangle = 0$ so that $\langle f, f\rangle = \langle f\circ T, f\rangle$. It follows that

$$\|f - f\circ T\|^2 = \langle f, f - f\circ T\rangle - \langle f\circ T, f\rangle + \langle f\circ T, f\circ T\rangle = -\|f\|^2 + \|f\|^2 = 0 ,$$

i.e., $f \in F$. □

Continuation of the proof of the ergodic theorem for d > 1:
For $f \in L^1_\mu$ let $\Delta f := A^+ f - A^- f$. We must show that $\Delta f = 0$ μ-a.s. To this end we decompose f into a sum of functions from either the space F or N of the previous lemma plus a small remainder term. Let $\epsilon > 0$.

1. $f = u_1 + r_0$ where $u_1 \in L^2_\mu$ and $\int |r_0|\, d\mu < \epsilon$. This is possible because L^2_μ is $\|\,.\,\|_1$-dense in L^1_μ.
2. $u_1 = f_1 + u_2 + \cdots + u_k + r_1$ for some $k \geq 2$ where $f_1 \in F$, $u_j = h_j - h_j \circ T^{g_j} \in N$, $g_j \in G$ ($j = 2, \ldots, k$) and $\int |r_1|\, d\mu \leq \|r_1\|_2 < \epsilon$, see Lemma 2.1.8.
3. $h_j = f_j + r_j$ where f_j is bounded and $\int |r_j|\, d\mu < \frac{\epsilon}{2k}$ ($j = 2, \ldots, k$). For example one can choose $f_j = h_j \cdot 1_{\{|h_j| \leq M\}}$ for some sufficiently large $M > 0$.

Let $R := r_0 + r_1 + \sum_{j=2}^k (r_j - r_j \circ T^{g_j})$. Then $\int |R|\, d\mu < 3\epsilon$ and

$$f = f_1 + \sum_{j=2}^{k} (f_j - f_j \circ T^{g_j}) + R$$

so that

$$\Delta f \leq \Delta f_1 + \sum_{j=2}^{k} \Delta(f_j - f_j \circ T^{g_j}) + \Delta R\,.$$

Now $\Delta f_1 = A^+ f_1 - A^- f_1 = f_1 - f_1 = 0$, and for $j = 2, \ldots, k$ we observe that

$$\begin{aligned}
|A_n(f_j - f_j \circ T^{g_j})| &= \left| \frac{1}{\lambda_n} \sum_{g \in \Lambda_n} f_j \circ T^g - \frac{1}{\lambda_n} \sum_{g \in \Lambda_n + g_j} f_j \circ T^g \right| \\
&\leq \frac{1}{\lambda_n} \sum_{g \in \Lambda_n \Delta (\Lambda_n + g_j)} |f_j \circ T^g| \\
&\leq \frac{1}{\lambda_n} \cdot 2d\, |g_j|\, n^{d-1} \cdot \|f_j\|_\infty \\
&= 2dn^{-1} |g_j|\, \|f_j\|_\infty\,,
\end{aligned}$$

so that $A^-(f_j - f_j \circ T^{g_j}) = A^+(f_j - f_j \circ T^{g_j}) = 0$ and hence $\Delta(f_j - f_j \circ T^{g_j}) = 0$. It follows that

$$0 \leq \Delta f \leq \Delta R = A^+ R - A^- R \leq 2A^+ |R|\,.$$

In view of (2.6) this implies

$$\int \Delta f\, d\mu \leq 2 \int A^+ |R|\, d\mu \leq 2\gamma_d \int |R|\, d\mu \leq 2\gamma_d \cdot 3\epsilon\,.$$

As $\epsilon > 0$ was arbitrary, we conclude that $\int \Delta f\, d\mu = 0$ and hence $\Delta f = 0$ μ-a.s. □

2.1.9 Remark The proof of the ergodic theorem given above goes back to articles by Kamae [30] and Katznelson and Weiss [32]. More traditional proofs, based on a so called maximal ergodic theorem, can be found in many textbooks on ergodic theory, in particular in Krengel's book [34] devoted completely to ergodic theorems. ◇

2.1.10 Exercise Give a short proof of the L^2*-ergodic theorem* due to von Neumann:

$$\lim_{n\to\infty}\left\|\lambda_n^{-1}\sum_{g\in\Lambda_n} f\circ T^g - \bar{f}\right\|_2 = 0 \quad \forall f\in L^2_\mu ,$$

where $\bar{f}$ is the orthogonal projection of f on the subspace F from Lemma 2.1.8. ◇

2.2 Ergodicity and mixing

The ergodic theorem is a rather general Law of Large Numbers for measure preserving dynamical systems, except that the limit $\bar{f}$ is not necessarily constant. Quite often, however, the limit turns out to be just $\mu(f)$ (as in Example 1.4.5). More generally, as $\bar{f}$ is a version of $E_\mu[f \mid \mathcal{I}_\mu(\mathcal{T})]$, the σ-algebra $\mathcal{I}_\mu(\mathcal{T})$ must be studied, and cases where it contains only null sets and sets of full measure are of particular interest, because in such instances $\bar{f}$ is constant μ-a.s.

2.2.1 Definition *The m.p.d.s.* $(X,\mathcal{B},\mu,\mathcal{T})$ *is ergodic if* $\mu(A)=0$ *or* $\mu(A)=1$ *for all* $A\in\mathcal{I}_\mu(\mathcal{T})$.

2.2.2 Lemma *For a m.p.d.s.* $(X,\mathcal{B},\mu,\mathcal{T})$ *the following statements are equivalent:*

1. $(X,\mathcal{B},\mu,\mathcal{T})$ *is ergodic.*
2. *For all* $f\in L^p_\mu$ $(1\le p\le\infty)$ *we have: if* $f\circ T^g = f \bmod \mu$ *for all* $g\in G$, *then* $f = \text{const} \bmod \mu$.
3. $\bar{f} = \int f\,d\mu \bmod \mu$ *for each measurable* $f: X\to\mathbb{R}$ *with* $f^+\in L^1_\mu$.
4. *For all* $A,B\in\mathcal{B}$

$$\lim_{n\to\infty}\frac{1}{\lambda_n}\sum_{g\in\Lambda_n}\mu(T^{-g}A\cap B) = \mu(A)\mu(B)\,. \tag{2.9}$$

5. *There is an* $\cap$*-stable family* $\mathcal{A}\subseteq\mathcal{B}$ *generating* $\mathcal{B}$ *such that (2.9) holds for all* $A,B\in\mathcal{A}$.
6. *If* ν *is another* $\mathcal{T}$*-invariant probability measure and if* $\nu\ll\mu$, *then* $\nu=\mu$.

Proof:
$1 \Rightarrow 3$: For $f, g : X \to [-\infty, \infty]$ let $f \vee g := \max\{f, g\}$. If $f \in L^1_\mu$, then simply apply the ergodic theorem. Otherwise $\int f\,d\mu = -\infty$ so that $\int f \vee r\,d\mu \to -\infty$ as $r \to -\infty$. As $f \vee r \in L^1_\mu$, the ergodic theorem implies

$$\bar{f} \le \overline{f \vee r} = \int f \vee r\,d\mu \to -\infty = \int f\,d\mu \quad \mu\text{-a.e.}$$

$3 \Rightarrow 2$: If $f \circ T^g = f \bmod \mu$ for all $g \in G$, then $f = \bar{f} = \int f\,d\mu = \text{const} \bmod \mu$.

$2 \Rightarrow 1$: If $A \in \mathcal{I}_\mu(\mathcal{T})$, then $1_A \circ T^g = 1_A \bmod \mu$ for all $g \in G$ so that $1_A = \text{const} \bmod \mu$. This implies $\mu(A) = 0$ or $\mu(A) = 1$.

$3 \Rightarrow 5$: Apply the L^1_μ-convergence in the ergodic theorem to $f = 1_A$.

$5 \Rightarrow 4$: For $A, B \in \mathcal{B}$ let

$$\begin{aligned} \mathcal{F}_A &:= \{B \in \mathcal{B} : (2.9) \text{ holds for } A \text{ and } B\}\,, \\ \mathcal{G}_B &:= \{A \in \mathcal{B} : (2.9) \text{ holds for } A \text{ and } B\}\,. \end{aligned}$$

$\mathcal{F}_A$ and $\mathcal{G}_B$ are Dynkin systems (see A.4.1) with $\mathcal{A} \subseteq \mathcal{F}_A$ and $\mathcal{A} \subseteq \mathcal{G}_B$ for $A, B \in \mathcal{A}$. We show this for $\mathcal{F}_A$, the proof for $\mathcal{G}_B$ is similar: Obviously, $A, X \in \mathcal{F}_A$, and if $B \in \mathcal{F}_A$ then also $X \setminus B \in \mathcal{F}_A$. So suppose that $B_1, B_2, B_3, \ldots \in \mathcal{F}_A$ are pairwise disjoint and let $B = \bigcup_{i=1}^\infty B_i$. Then

$$\begin{aligned} \mu(A)\mu(B) &= \sup_{k>0} \mu(A) \cdot \mu\left(\bigcup_{i=1}^k B_i\right) \\ &= \sup_{k>0} \lim_{n\to\infty} \frac{1}{\lambda_n} \sum_{g\in\Lambda_n} \mu\left(T^{-g}A \cap \bigcup_{i=1}^k B_i\right) \\ &\le \lim_{n\to\infty} \frac{1}{\lambda_n} \sum_{g\in\Lambda_n} \mu\left(T^{-g}A \cap B\right) \\ &\le \inf_{k>0}\left(\lim_{n\to\infty} \frac{1}{\lambda_n} \sum_{g\in\Lambda_n} \mu\left(T^{-g}A \cap \bigcup_{i=1}^k B_i\right) + \mu\left(\bigcup_{i=k+1}^\infty B_i\right)\right) \\ &= \inf_{k>0}\left(\mu(A) \cdot \mu\left(\bigcup_{i=1}^k B_i\right) + \mu\left(\bigcup_{i=k+1}^\infty B_i\right)\right) \\ &\le \mu(A)\mu(B)\,. \end{aligned}$$

As $\mathcal{A}$ is an $\cap$-stable generator of the σ-algebra $\mathcal{B}$, it follows that $\mathcal{F}_A = \mathcal{G}_B = \mathcal{B}$ for all $A, B \in \mathcal{A}$ (see A.4.2). Therefore $\mathcal{A} \subseteq \mathcal{F}_A$ and $\mathcal{A} \subseteq \mathcal{G}_B$ for all $A, B \in \mathcal{B}$, and the same argument as before yields $\mathcal{F}_A = \mathcal{G}_B = \mathcal{B}$ for all $A, B \in \mathcal{B}$, i.e., (2.9) holds for all $A, B \in \mathcal{B}$.

$4 \Rightarrow 1$: If $A \in \mathcal{I}_\mu(\mathcal{T})$, then $\mu(T^{-g}A \cap A) = \mu(A)$ for all $g \in G$ so that (2.9) applied to $A = B$ yields $\mu(A) = \mu(A)^2$, i.e., $\mu(A) = 0$ or $\mu(A) = 1$. Hence $\mathcal{T}$ is ergodic.

$3 \Rightarrow 6$: Suppose ν is $\mathcal{T}$-invariant and $\nu \ll \mu$. By assertion 3 we have $\overline{1_A} = \mu(A)$ μ-a.s. and hence also ν-a.s. for any $A \in \mathcal{B}$. On the other hand, the ergodic theorem applied to $(X, \mathcal{B}, \nu, \mathcal{T})$ yields $\int \overline{1_A}\, d\nu = \nu(A)$. Therefore $\mu(A) = \nu(A)$ for all $A \in \mathcal{B}$.

$6 \Rightarrow 1$: For $A \in \mathcal{I}_\mu(\mathcal{T})$ with $\mu(A) > 0$ define a probability measure μ_A on $\mathcal{B}$ by $\mu_A(B) = \mu(A \cap B)/\mu(A)$. Then $\mu_A \ll \mu$ and μ_A is evidently $\mathcal{T}$-invariant. Hence $\mu_A = \mu$, i.e., $\mu(A) = 1$. □

2.2.3 Remark In his 1884 publication [8], p.79, Boltzmann uses for the first time the term *Ergode* to denote an energy surface in the phase space of a mechanical system on which no further integrals of motion exist. He had the idea that a trajectory of the system inside an *Ergode* goes through all its points, and this served him to justify his hypothesis that the time average of a measured quantity along a trajectory equals the phase space average of this quantity over the *Ergode*. This is exactly how statement 3 of Lemma 2.2.2 characterizes an ergodic m.p.d.s. So, in modern terminology, an *Ergode* is an ergodic subsystem of a m.p.d.s. In the next section we see that each m.p.d.s. can be decomposed into ergodic components.

Today it seems obvious to us that Boltzmann's idea of a trajectory passing through all points of an *Ergode* is hopelessly wrong. The more remarkable it seems that he guessed the assertion of the ergodic theorem more or less correctly. In any case his picture of an *Ergode* may explain the etymological roots of this term as in [18, Remark 88]: it is made up of the Greek words $\grave{\varepsilon}\rho\gamma o\nu$ = energy and $\acute{o}\delta\grave{o}\varsigma$ = path. An alternative etymological explanation is advocated in [39]. ◇

2.2.4 Exercise Let $(X, \mathcal{B}, \mu, \mathcal{T})$ be an ergodic m.p.d.s. Suppose that $\mu\{x\} > 0$ for some $x \in X$. Prove that μ is concentrated on a periodic orbit of $\mathcal{T}$, i.e., there are $n_1, \ldots, n_d \in \mathbb{Z}_+$ such that $T^{n_i e_i}x = x$ for $i = 1, \ldots, d$ and μ is the equidistribution on $\{T^g x : g \in G, 0 \le g_i < n_i\ (i = 1, \ldots, d)\}$. ◇

A much stronger property than ergodicity is the kind of asymptotic independence of observations defined in

2.2.5 Definition *The m.p.d.s.* $(X, \mathcal{B}, \mu, \mathcal{T})$ *is mixing, if for all* $A, B \in \mathcal{B}$ *the following holds:*

$$\forall \epsilon > 0\ \exists n_0 > 0\ \forall g \in G \setminus \Lambda_{n_0} : \ |\mu(T^{-g}A \cap B) - \mu(A) \cdot \mu(B)| < \epsilon\,.$$

We abbreviate this as

$$\lim_{g \in G} \mu(T^{-g}A \cap B) = \mu(A) \cdot \mu(B)\,. \tag{2.10}$$

2.2.6 Corollary *An immediate consequence of Lemma 2.2.2 is that mixing implies ergodicity.*

2.2.7 Lemma *For a m.p.d.s. $(X, \mathcal{B}, \mu, \mathcal{T})$ the following statements are equivalent:*

1. *$(X, \mathcal{B}, \mu, \mathcal{T})$ is mixing.*
2. *There is an $\cap$-stable family $\mathcal{A} \subseteq \mathcal{B}$ generating $\mathcal{B}$ such that (2.10) holds for all $A, B \in \mathcal{A}$.*
3. *For all $f, h \in L^2_\mu$*
$$\lim_{g \in G} \int (f \circ T^g) \cdot h \, d\mu = \int f \, d\mu \cdot \int h \, d\mu . \tag{2.11}$$
4. *For all $f \in L^2_\mu$*
$$\lim_{g \in G} \int (f \circ T^g) \cdot f \, d\mu = \left(\int f \, d\mu \right)^2 .$$

So mixing is equivalent to L^2_μ-weak convergence of $f \circ T^g$ to $\int f \, d\mu$ and to asymptotic vanishing of autocorrelations. In contrast to that, ergodicity implies these properties only for the corresponding averaged quantities.

Proof:

$1 \Rightarrow 2$: Trivial.

$2 \Rightarrow 1$: Just as "$5 \Rightarrow 4$" in the previous lemma.

$1 \Leftrightarrow 3$: If f, h are indicator functions of measurable sets, property (2.11) is just the mixing property. By linearity of the integral this identity extends to finite linear combinations of indicator functions, and as these functions are dense in L^2_μ, it extends to arbitrary $f, h \in L^2_\mu$.

$3 \Rightarrow 4$: Trivial.

$4 \Rightarrow 3$: For $f \in L^2_\mu$ let
$$\mathcal{V}_f := \{h \in L^2_\mu : \text{(2.11) holds for } f \text{ and } h\} .$$

$\mathcal{V}_f$ is a closed linear subspace of L^2_μ which contains f (by assumption) and the constant functions (trivially). Furthermore, by $\mathcal{T}$-invariance of μ, also $f \circ T^{g'} \in \mathcal{V}_f$ for all $g' \in G$. We must show that $\mathcal{V}_f = L^2_\mu$, i.e., $\mathcal{V}_f^\perp = \{0\}$. So let $h \in \mathcal{V}_f^\perp$. Then $\int (f \circ T^g) \cdot h \, d\mu = 0$ for all $g \in G$ and also $\int h \, d\mu = 0$, so that $h \in \mathcal{V}_f$. As $h \in \mathcal{V}_f^\perp$ this implies $h = 0$. □

2.2.8 Example (Circle rotation) Let $X = S^1 := \{z \in \mathbb{C} : |z| = 1\}$, and for $a \in [0,1)$ let $T_a : X \to X$, $T_a z = e^{2\pi i a} z$ denote the rotation by $2\pi a$. T_a leaves the normalized one-dimensional Lebesgue measure μ on X invariant. In order to examine ergodic and mixing properties of T_a, functions $f \in L^2_\mu$ are represented by their Fourier series $f(z) = \sum_{n\in\mathbb{Z}} f_n \cdot z^n$. Then $f(T_a^k z) = f(e^{2k\pi i a} z) = \sum_{n\in\mathbb{Z}} (f_n e^{2kn\pi i a}) \cdot z^n$ for each $k \in \mathbb{Z}$, and the T_a-invariance of f is equivalent to $f_n \cdot (1 - e^{2n\pi i a}) = 0$ for all $n \in \mathbb{Z}$. Consequently

$$\begin{aligned}
& T_a \text{ is ergodic} \\
\Leftrightarrow\ & \text{all } T_a\text{-invariant } L^2_\mu\text{-functions are constant mod } \mu \\
\Leftrightarrow\ & e^{2n\pi i a} \neq 1 \text{ for all } n \in \mathbb{Z} \setminus \{0\} \\
\Leftrightarrow\ & a \notin \mathbb{Q}\,.
\end{aligned}$$

In particular, T_a is not mixing for rational a. That T_a is never mixing follows from looking at the function $f(z) = z + z^{-1}$, for which $\int f\, d\mu = f_0 = 0$ but

$$\begin{aligned}
\int (f \circ T_a^k) \cdot f\, d\mu &= \int (e^{2k\pi i a} z + e^{-2k\pi i a} z^{-1})(z + z^{-1})\, d\mu \\
&= e^{2k\pi i a} + e^{-2k\pi i a} \not\to 0
\end{aligned}$$

as $k \to \infty$ for irrational a. ◇

2.2.9 Exercise Show that an irrational rotation T_a is uniquely ergodic, i.e., that the normalized Lebesgue measure is its only invariant measure. *Hint:* Use Exercise 1.4.4. ◇

2.2.10 Example (Bernoulli shift) As in Section 1.2 let Σ be a finite set, $\Omega = \Sigma^G$, $G = \mathbb{Z}^d$ or $G = \mathbb{Z}^d_+$, and let $T^g : \Omega \to \Omega$ denote the shift by g. Let $\mathcal{C}_n$ be the family of cylinder sets that depend only on coordinates $g \in \Lambda_n$, i.e.,

$$\mathcal{C}_n := \{[\sigma_g]_{g\in\Lambda_n} : \sigma_g \in \Sigma\ (g \in \Lambda_n)\} \tag{2.12}$$

where

$$[\sigma_g]_{g\in\Lambda_n} := \{\omega \in \Omega : \omega_g = \sigma_g\ (g \in \Lambda_n)\}\,,$$

and set $\mathcal{C} := \bigcup_n \mathcal{C}_n$. $\mathcal{C}$ is an $\cap$-stable generator for the canonical σ-algebra $\mathcal{B}$ on Ω. (see A.4.4).

We fix a probability vector $q = (q_\sigma : \sigma \in \Sigma)$ on Σ and denote the corresponding Bernoulli measure on Ω by $\mu := q^{\times G}$. For all $A, B \in \mathcal{C}$ there is $k > 0$ such that $T^{-g}A$ and B are specified on mutually disjoint sets of coordinates and hence are independent events under μ if $|g_i| > k$ for at least one index i. It follows from Lemma 2.2.7 and Corollary 2.2.6 that the dynamical system $(\Omega, \mathcal{B}, \mu, (T^g : g \in G))$ (called a *Bernoulli shift*) is mixing and, a fortiori, ergodic. ◇

2.2.11 Remark There is another intermediate notion of mixing between ergodicity and mixing: the m.p.d.s. $(X, \mathcal{B}, \mu, \mathcal{T})$ is *weakly mixing*, if

$$\lim_{n\to\infty} \frac{1}{\lambda_n} \sum_{g\in\Lambda_n} |\mu(T^{-g}A \cap B) - \mu(A)\mu(B)| = 0 \quad \text{for all } A, B \in \mathcal{B}.$$

Obviously mixing implies weak mixing which in turn implies ergodicity. ◇

2.2.12 Exercise Prove that the m.p.d.s. $(X, \mathcal{B}, \mu, \mathcal{T})$ is weakly mixing if and only if for all $A, B \in \mathcal{B}$

$$\lim_{n\to\infty} \frac{1}{\lambda_n} \sum_{g\in\Lambda_n} \left(\mu(T^{-g}A \cap B) - \mu(A)\mu(B)\right)^2 = 0 .$$

◇

For the next exercise we need the notion of a *product system*. The product of two m.p.d.s. $(X_i, \mathcal{B}_i, \mu_i, \mathcal{T}_i)$ $(i = 1, 2)$ is the m.p.d.s. $(X_1 \times X_2, \mathcal{B}_1 \otimes \mathcal{B}_2, \mu_1 \times \mu_2, \mathcal{T}_1 \times \mathcal{T}_2)$. Here $(X_1 \times X_2, \mathcal{B}_1 \otimes \mathcal{B}_2, \mu_1 \times \mu_2)$ is simply the product probability space, and $\mathcal{T}_1 \times \mathcal{T}_2 = (T_1^{g_1} \times T_2^{g_2} : g_1 \in G_1, g_2 \in G_2)$. Of course $(g_1, g_2) \mapsto T_1^{g_1} \times T_2^{g_2}$ is an action of the (semi)group $G_1 \times G_2$ on $X_1 \times X_2$. It preserves the measure $\mu_1 \times \mu_2$, because this measure is obviously preserved for sets from the $\cap$-stable generator $\{B_1 \times B_2 : B_1 \in \mathcal{B}_1, B_2 \in \mathcal{B}_2\}$ of $\mathcal{B}_1 \times \mathcal{B}_2$.

2.2.13 Exercise Prove that the following statements are equivalent for a m.p.d.s. $(X, \mathcal{B}, \mu, \mathcal{T})$:

1. $(X, \mathcal{B}, \mu, \mathcal{T})$ is weakly mixing.
2. $(X \times X, \mathcal{B} \otimes \mathcal{B}, \mu \times \mu, \mathcal{T} \times \mathcal{T})$ is ergodic.
3. $(X \times X, \mathcal{B} \otimes \mathcal{B}, \mu \times \mu, \mathcal{T} \times \mathcal{T})$ is weakly mixing.

Hint: I suggest proving $1 \Rightarrow 3 \Rightarrow 2 \Rightarrow 1$ and using Exercise 2.2.12 for the last step. ◇

2.2.14 Exercise Prove that irrational rotations are not weakly mixing.
Hint: There are many ways to prove this. One might check that if $A = B$ is an interval of measure $\frac{1}{4}$, then the definition of weak mixing is violated. Another possibility is to use Exercise 2.2.13. ◇

2.3 The ergodic decomposition

If the m.p.d.s. $(X, \mathcal{B}, \mu, \mathcal{T})$ is not ergodic, then there exists a set $A \in \mathcal{I}(\mathcal{T})$ with $0 < \mu(A) < 1$, and the system decomposes into two subsystems $(A, \mathcal{B}_A, \mu_A, \mathcal{T}_A)$

and $(A^c, \mathcal{B}_{A^c}, \mu_{A^c}, \mathcal{T}_{A^c})$ that can be considered independently. Here $A^c := X \setminus A$, and $\mathcal{B}_A := \{B \cap A : B \in \mathcal{B}\}$ and $\mathcal{T}_A := (T^g_{|A} : g \in G)$ are the restrictions of $\mathcal{B}$ and $\mathcal{T}$ to A. By μ_A we denote the normalization of the restriction of μ to A, i.e., $\mu_A(B) := \mu(A \cap B)/\mu(A)$.

From a purely measure theoretic point of view nothing is changed if instead of the two restricted subsystems the m.p.d.s.'s $(X, \mathcal{B}, \mu_A, \mathcal{T})$ and $(X, \mathcal{B}, \mu_{A^c}, \mathcal{T})$ are studied, because their phase spaces differ only by sets of measure 0 from those of the original decomposition. Because of that (and in order to keep the technical complications limited) we are going to decompose only the invariant measures into ergodic components and do not attempt to obtain a corresponding disjoint decomposition of the underlying measurable space, which the reader can find, e.g., in [51]. An approach based on properties of the weak topology on the space of probability measures is used in [25].

If $\mathcal{I}(\mathcal{T})$ is generated by finitely or countably many atoms (mod μ), a corresponding finite or countable decomposition into ergodic components is obtained by the above reasoning. Otherwise no reduction to ergodic components of positive measure is possible. Instead, the conditional probabilities $\mu(\,.\mid \mathcal{I}_\mu(\mathcal{T}))$ and their representations by conditional probability distributions (see A.4.40) play a key role.

2.3.1 Theorem (Conditional probability distributions)
(See A.4.41.) Suppose X is a complete, separable metric space with Borel σ-algebra $\mathcal{B}$, μ is a Borel measure on X, and $\mathcal{C}$ is a sub-σ-algebra of $\mathcal{B}$. Then there is a conditional probability distribution $(\mu_x \mid x \in X)$ for $\mu(\,.\mid \mathcal{C})(x)$. If $(\tilde{\mu}_x \mid x \in X)$ is another conditional probability distribution for $\mu(\,.\mid \mathcal{C})(x)$, then $\mu_x = \tilde{\mu}_x$ for μ-a.e. x. We use the notation $\mu = \int \mu_x \, d\mu(x)$ for this fact.

2.3.2 Remark Let $\mathcal{C} = \mathcal{I}_\mu(\mathcal{T})$ and $A \in \mathcal{B}$. As the ergodic limit $\overline{1_A}$ of 1_A is a version of $\mu(A \mid \mathcal{I}_\mu(\mathcal{T}))$ (see Remark 2.1.6), the equality $\overline{1_A}(x) = \mu_x(A)$ holds for μ-a.e. $x \in X$. ◇

For a m.p.d.s. Theorem 2.3.1 can be extended to

2.3.3 Theorem (Ergodic decomposition)
Let $(X, \mathcal{B}, \mu, \mathcal{T})$ be a m.p.d.s. with underlying probability space $(X, \mathcal{B}, \mu)$ as in Theorem 2.3.1. Let $\mathcal{C} = \mathcal{I}_\mu(\mathcal{T})$. Then there is a conditional probability distribution $(\mu_x \mid x \in X)$ for $\mu(\,.\mid \mathcal{C})(x)$ such that

1. *$\mu_{T^g x} = \mu_x$ for all $x \in X$ and $g \in G$, and*
2. *$(X, \mathcal{B}, \mu_x, \mathcal{T})$ is an ergodic m.p.d.s. for each $x \in X$.*

Proof: Let $(\mu_x \mid x \in X)$ be a conditional probability distribution for $\mu(\,.\mid \mathcal{C})(x)$ from Theorem 2.3.1. For each $A \in \mathcal{B}$ the function $x \mapsto \mu_x(A)$ is $\mathcal{C}$-measurable, and since $\mathcal{C}$ is a $\mathcal{T}$-invariant σ-algebra, the functions $x \mapsto \mu_{T^g x}(A)$ are also $\mathcal{C}$-measurable ($g \in G$). As the $\mathcal{T}$-invariance of the measure μ implies

$$\int_C \mu_{T^g x}(A)\, d\mu(x) = \int_{T^{-g}C} \mu_{T^g x}(A)\, d\mu(x) = \int_C \mu_x(A)\, d\mu(x)$$

for all $A \in \mathcal{B}$ and $C \in \mathcal{C}$, and as $\mathcal{B}$ is countably generated (see A.4.3), $\mu_{T^g x} = \mu_x$ ($g \in G$) for μ-a.e. x by the uniqueness assertion of Theorem 2.3.1.

We prove the $\mathcal{T}$-invariance of the μ_x. To this end fix $g \in G$ and let $A \in \mathcal{B}$ and $C \in \mathcal{C}$. By the defining properties of conditional probability distributions

$$\int_C \mu_x(T^{-g}A)\, d\mu(x) = \mu(C \cap T^{-g}A) = \mu(T^{-g}(C \cap A)) = \mu(C \cap A)\,.$$

Therefore $(T^g\mu_x \mid x \in X)$ is also a conditional probability distribution for $\mu(\,.\mid \mathcal{C})(x)$, such that $T^g\mu_x = \mu_x$ for μ-a.e. x by Theorem 2.3.1.

In the next step we prove ergodicity of the μ_x. The basic idea is simple: For $C \in \mathcal{C}$ the equality $\mu_x(C) = \overline{1_C}(x) = 1_C(x) = \delta_x(C) \in \{0,1\}$ holds μ-a.s., see Remark 2.3.2. If we knew that $\mathcal{C} = \mathcal{I}_\mu(\mathcal{T})$ had an $\cap$-stable generator, we could conclude that $\mu_x|\mathcal{C} = \delta_x|\mathcal{C}$ for μ-a.e. x and hence that μ_x is ergodic for μ-a.e. x. Unfortunately the construction of such a generator is technically rather demanding. Therefore we choose a simpler, though less appealing proof: Let $A, B \in \mathcal{B}$. Because of Remark 2.1.6 and of Remark 2.3.2 applied to the indicator 1_B,

$$\begin{aligned}\int_B \overline{1_A}(y)\, d\mu_x(y) &= E_\mu[\overline{1_A} \cdot 1_B \mid \mathcal{I}_\mu(\mathcal{T})](x) = \overline{1_A}(x) \cdot E_\mu[1_B \mid \mathcal{I}_\mu(\mathcal{T})](x) \\ &= \mu_x(A) \cdot \mu_x(B) \quad \text{for } \mu\text{-a.e. } x.\end{aligned}$$

For such x we apply the ergodic theorem to the m.p.d.s. $(X, \mathcal{B}, \mu_x, T)$ and conclude that

$$\begin{aligned}&\lim_{n\to\infty} \frac{1}{\lambda_n} \sum_{g \in \Lambda_n} \mu_x(T^{-g}A \cap B) \\ &= \lim_{n\to\infty} \int_B \frac{1}{\lambda_n} \sum_{g \in \Lambda_n} 1_A \circ T^g\, d\mu_x \\ &= \int_B \overline{1_A}\, d\mu_x = \mu_x(A) \cdot \mu_x(B)\,. \qquad (2.13)\end{aligned}$$

Now let $\mathcal{A}$ be a countable $\cap$-stable generator for $\mathcal{B}$. (Such a generator exists because of the separability assumption on $\mathcal{B}$, see A.4.3.) Then there is a μ-null set $N \in \mathcal{B}$ such that (2.13) holds for all $x \in X \setminus N$ and all $A, B \in \mathcal{A}$, and because of Lemma 2.2.2, μ_x is ergodic for $x \in X \setminus N$.

In a final step we modify $(\mu_x \mid x \in X)$ for x in a μ-null set such that the properties claimed in the theorem hold for all $x \in X$. To this end let $Y_0 \subseteq X$ be a set of full μ-measure such that these properties hold for all $x \in Y_0$. Let $Y := \bigcap_{g\in G} T^{-g}Y_0$. Then $\mu(Y) = 1$ and $T^g(Y) \subseteq Y$ $(g \in G)$. By definition of Y we have: if $x, y \in Y$ and if there are $g, g' \in G$ such that $T^g x = T^{g'} y$, then $\mu_x = \mu_y$. Therefore, for $x \in \bigcup_{g\in G} T^{-g}Y$, a measure $\tilde{\mu}_x$ can be unambiguously defined by $\tilde{\mu}_x := \mu_{T^g x}$ for any g with $T^g x \in Y$. (Observe that for all $g, g' \in G$ there are $h, h' \in G$ such that $g + h = g' + h'$!) As $T^g x \in Y \subseteq Y_0$, the measure $\tilde{\mu}_x = \mu_{T^g x}$ is $\mathcal{T}$-invariant and ergodic. For the remaining x in the $\mathcal{T}$-invariant set $X \setminus \bigcup_{g\in G} T^{-g}Y$ let $\tilde{\mu}_x = \mu_{x_0}$, where x_0 is an arbitrarily chosen point in Y. As $\tilde{\mu}_x = \mu_x$ for all x in the full measure set Y, $(\tilde{\mu}_x \mid x \in X)$ is again a conditional probability distribution for $\mu(\,.\mid \mathcal{C})(x)$. By construction, all $\tilde{\mu}_x$ are $\mathcal{T}$-invariant ergodic measures and satisfy $\tilde{\mu}_{T^g x} = \tilde{\mu}_x$ for all $g \in G$ and $x \in X$. □

2.3.4 Exercise As in Example 2.2.8 denote by T_a the rotation on S^1 by $2\pi a$ and let μ be the normalized Lebesgue measure on S^1. Determine the ergodic decomposition of

a) $(S^1, \mathcal{B}, \mu, (T_a^n : n \in \mathbb{Z}))$ for rational a,

b) $(S^1 \times S^1, \mathcal{B} \times \mathcal{B}, \mu \times \mu, ((T_a \times T_a)^n : n \in \mathbb{Z}))$ for both rational and irrational a. (See also Exercise 2.2.13.)

What about $T_a \times T_b$ for $a, b \in [0, 1)$? ◇

2.4 Return times and return maps

In order to analyse a given m.p.d.s. it is sometimes useful to relate its properties to those of another system which can be dealt with more easily. In this section we discuss a construction that can be applied to $\mathbb{Z}_+$-actions and that yields in some cases a system with more desirable properties. Concrete examples are treated in Chapter 6.

Let $(X, \mathcal{B}, \mu, T)$ be an ergodic m.p.d.s. Fix $A \in \mathcal{B}$ with $\mu(A) > 0$ and let $A_{ret} := \{x \in A : T^k x \in A \text{ for infinitely many } k\}$. Then $\mu(A \setminus A_{ret}) = 0$ by Birkhoff's ergodic theorem, and for each $x \in A_{ret}$ there exists $k > 0$ such that $T^k x \in A_{ret}$. Hence, replacing A by A_{ret} we may assume that the following *first return time* to A is well defined for every $x \in A$:

$$\tau_A(x) := \min\{k \geq 1 : T^k x \in A\}\,.$$

We use it to define the *first return map* on A (also called the *induced map* on A),

$$T_A : A \to A, \quad T_A(x) := T^{\tau_A(x)}(x)\,.$$

Observe that the ergodic measure μ was only used to guarantee that $\mu(A \setminus A_{ret}) = 0$. If we simply assume that $A = A_{ret}$, then T_A and τ_A are well defined without reference to any measure.

2.4.1 Theorem (Kac's recurrence theorem)
Suppose that $A_{ret} = A$ and that $(X, \mathcal{B}, \mu, T)$ is ergodic. Then τ_A and T_A have the following properties:

a) $\int_A \tau_A \, d\mu = 1$.
b) $\mu(T_A^{-1} B) = \mu(B)$ *for each measurable set* $B \subseteq A$.
c) *For* $n \geq 0$ *let* $\tau^{(n)} = \tau_A + \tau_A \circ T_A + \ldots + \tau_A \circ T_A^{n-1}$. *($\tau^{(n)}(x)$ is the time of the n-th return of x to A.) Then* $\lim_{n\to\infty} n^{-1} \tau^{(n)}(x) = \mu(A)^{-1}$ *for μ-a.e. x.*

Proof: Let $x \in A$. Then $T^k x \in A$ if and only if $k = \tau^{(j)}(x)$ for some $j \geq 0$. In this case $T^k x = T_A^j x$. Consider a function $f : A \to \mathbb{R}$.

$$(1_A \cdot f)(T^k x) = \begin{cases} f(T_A^j x) & \text{if } k = \tau^{(j)}(x), \\ 0 & \text{otherwise.} \end{cases} \tag{2.14}$$

We apply this identity to $f = \tau_A$. In view of Birkhoff's ergodic theorem and the ergodicity of μ it follows that for μ-a.e. $x \in A$

$$\int_A \tau_A \, d\mu = \lim_{n\to\infty} \frac{1}{n} \sum_{k=0}^{n-1} (1_A \cdot \tau_A)(T^k x) = \lim_{l\to\infty} \frac{1}{\tau^{(l)}(x)} \sum_{j=0}^{l-1} \tau_A(T_A^j x) = 1\,.$$

This proves the first assertion of the theorem. The third one follows in the same way from (2.14) with $f = 1_A$:

$$\mu(A) = \lim_{n\to\infty} \frac{1}{n} \sum_{k=0}^{n-1} 1_A(T^k x) = \lim_{l\to\infty} \frac{1}{\tau^{(l)}(x)} \sum_{j=0}^{l-1} 1_A(T_A^j x) = \lim_{l\to\infty} \frac{l}{\tau^{(l)}(x)}$$

for μ-a.e. $x \in A$.

In order to show the second assertion, we let $B \subseteq A$ and conclude from Birkhoff's ergodic theorem and (2.14) for $f = 1_B \circ T_A$ that

$$\begin{aligned} \mu(A \cap T_A^{-1} B) &= \lim_{n\to\infty} \frac{1}{n} \sum_{k=0}^{n-1} ((1_B \circ T_A) \cdot 1_A)(T^k x) \\ &= \lim_{l\to\infty} \frac{1}{\tau^{(l)}(x)} \sum_{j=0}^{l-1} 1_B(T_A^{j+1} x) = \lim_{l\to\infty} \frac{\tau^{(l+1)}(x)}{\tau^{(l)}(x)} \cdot \frac{1}{\tau^{(l+1)}(x)} \sum_{j=1}^{l} 1_B(T_A^j x) \\ &= \lim_{l\to\infty} \frac{1}{\tau^{(l+1)}(x)} \sum_{j=0}^{l} 1_B(T_A^j x) \quad \text{for } \mu\text{-a.e. } x \end{aligned}$$

where we used the third assertion for the last equality. Applying Birkhoff's ergodic theorem and (2.14) for $f = 1_B$ once more it follows that $\mu(A \cap T_A^{-1}B) = \mu(A \cap B) = \mu(B)$. □

This theorem has the following immediate corollary:

2.4.2 Corollary *Denote by* μ_A *the normalized restriction of* μ *to* A *and let* $\mathcal{B}_A = \{B \cap A : B \in \mathcal{B}\}$. *Then* $(A, \mathcal{B}_A, \mu_A, T_A)$ *is a m.p.d.s., called the induced system.*

2.4.3 Exercise Prove that $(A, \mathcal{B}_A, \mu_A, T_A)$ is also ergodic. ◇

2.4.4 Exercise Use ergodic decompositions to formulate and prove a version of Kac's recurrence theorem for nonergodic systems $(X, \mathcal{B}, \mu, T)$. ◇

2.4.5 Exercise Let T be a measurable transformation of the measurable space $(X, \mathcal{B})$. Suppose that $A = A_{ret}$ for some $A \in \mathcal{B}$ so that T_A and τ_A are well defined. Prove that each ergodic T_A-invariant probability measure ν on $(A, \mathcal{B}_A)$ with $\int \tau_A \, d\nu < \infty$ gives rise to an ergodic T-invariant probability measure μ on $(X, \mathcal{B})$ such that $\nu = \mu_A$ and $\int f \, d\mu = \int_A \sum_{k=0}^{\tau_A - 1} f \circ T^k \, d\nu$. ◇

2.4.6 Remark There is also a more elementary proof of Kac's recurrence theorem that does not make use of Birkhoff's ergodic theorem, see e.g., [46], [34]. ◇

2.5 Factors and extensions

As in practically all branches of modern mathematics, there is a notion of homomorphism between m.p.d.s.'s. Therefore the usual constructions like factors and extensions of systems can be performed and conditions under which systems are isomorphic can be studied. As this structural reasoning plays only a marginal role in the further chapters of this text, we mention these concepts here only briefly.

Let $(X_i, \mathcal{B}_i, \mu_i, \mathcal{T}_i)$ $(i = 1, 2)$ be two m.p.d.s.'s, both $\mathcal{T}_i$ acting under the same semigroup G. A $\mathcal{B}_1$-$\mathcal{B}_2$-measurable map $\Phi : X_1 \to X_2$ is a *homomorphism* from the first to the second system, if

$$\mu_1 \circ \Phi^{-1} = \mu_2 \quad \text{and} \quad \Phi \circ T_1^g = T_2^g \circ \Phi \text{ for all } g \in G.$$

As sets of measure zero do not matter in the study of m.p.d.s.'s, it suffices that Φ is defined μ_1-a.e. and that $\Phi \circ T_1^g = T_2^g \circ \Phi$ holds μ_1-a.e. For the same reason we may call $(X_2, \mathcal{B}_2, \mu_2, \mathcal{T}_2)$ a *factor* of $(X_1, \mathcal{B}_1, \mu_1, \mathcal{T}_1)$ and $(X_1, \mathcal{B}_1, \mu_1, \mathcal{T}_1)$ an *extension* of $(X_2, \mathcal{B}_2, \mu_2, \mathcal{T}_2)$, for if $A \in \mathcal{B}_2$ is disjoint from $\Phi(X_1)$, then $\mu_2(A) = \mu_1(\Phi^{-1}A) = 0$.

Suppose that $(X_1, \mathcal{B}_1, \mu_1, \mathcal{T}_1)$ is an extension of $(X_2, \mathcal{B}_2, \mu_2, \mathcal{T}_2)$ (both considered as $\mathbb{Z}_+^d$-actions). Suppose also that all $T \in \mathcal{T}_1$ are invertible so that $\mathcal{T}_1$ is actually a $\mathbb{Z}^d$-action. Consider the sub-σ-algebra $\mathcal{B}_0 = \sigma(\bigcup_{g \in \mathbb{Z}^d} T^{-g}\Phi^{-1}\mathcal{B}_2)$ of $\mathcal{B}_1$.

By construction, the action T_1 is $\mathcal{B}_0$-$\mathcal{B}_0$-measurable and Φ is $\mathcal{B}_0$-$\mathcal{B}_2$-measurable. Hence, $(X_1, \mathcal{B}_0, \mu_1|_{\mathcal{B}_0}, T_1)$ is also an extension of $(X_2, \mathcal{B}_2, \mu_2, T_2)$ by an invertible action, and, by construction again, $\mathcal{B}_0$ is the smallest sub-σ-algebra of $\mathcal{B}_1$ with this property. Therefore, the system $(X_1, \mathcal{B}_1, \mu_1, T_1)$ is called a *natural extension* of $(X_2, \mathcal{B}_2, \mu_2, T_2)$ if $\mathcal{B}_0 = \mathcal{B}_1 \bmod \mu_1$. One can prove that natural extensions always exist and that they are uniquely determined up to isomorphism. A complete proof of this assertion is difficult to locate in the literature.

If $(X_2, \mathcal{B}_2, \mu_2, T_2)$ is a factor of $(X_1, \mathcal{B}_1, \mu_1, T_1)$ and if $(X_1, \mathcal{B}_1, \mu_1, T_1)$ is ergodic (weakly mixing, mixing), then $(X_2, \mathcal{B}_2, \mu_2, T_2)$ is also ergodic (weakly mixing, mixing), see Exercise 2.5.4.

2.5.1 Example (Natural extensions for shift spaces) The shift spaces from Example 2.2.10 have particularly simple natural extensions. Let $\Omega_1 = \Sigma^{\mathbb{Z}^d}$, $\Omega_2 = \Sigma^{\mathbb{Z}^d_+}$, and denote by $\mathcal{B}_1, \mathcal{B}_2$ the corresponding Borel σ-algebras and by T_1, T_2 the (semi)groups of shift transformations on Ω_1 and Ω_2 respectively. Define $\Phi : \Omega_1 \to \Omega_2$ by $\Phi((\omega_g)_{g\in\mathbb{Z}^d}) = (\omega_g)_{g\in\mathbb{Z}^d_+}$. So Φ describes the restriction of a $\mathbb{Z}^d$ configuration to $\mathbb{Z}^d_+$. Obviously $\Phi \circ T_1^g = T_2^g \circ \Phi$ for all $g \in \mathbb{Z}^d_+$.

▷ If μ_1 is a T_1-invariant measure on Ω_1, then $\mu_1 \circ \Phi^{-1}$ is a T_2-invariant measure on Ω_2, because for each cylinder set $[\sigma_g]_{g\in\Lambda_n^+} \subset \Omega_2$ and $g' \in \mathbb{Z}^d_+$ we have:

$$\mu_1 \circ \Phi^{-1}(T_2^{-g'}[\sigma_g]_{g\in\Lambda_n^+}) = \mu_1 \circ T_1^{-g'}(\Phi^{-1}[\sigma_g]_{g\in\Lambda_n^+}) = \mu_1(\Phi^{-1}[\sigma_g]_{g\in\Lambda_n^+}) .$$

Therefore $(\Omega_2, \mathcal{B}_2, \mu_1 \circ \Phi^{-1}, T_2)$ is a factor of $(\Omega_1, \mathcal{B}_1, \mu_1, T_1)$.

▷ If μ_2 is a T_2-invariant measure on Ω_2, we can define a measure μ_1 on Ω_1 as follows: Let $[\sigma_g]_{g\in\Lambda_n}$ be a cylinder set in Ω_1. Take any $g' \in \mathbb{Z}^d_+$ such that $g' + \Lambda_n \subset \mathbb{Z}^d_+$ and define

$$\mu_1([\sigma_g]_{g\in\Lambda_n}) := \mu_2([\sigma_g]_{g\in g'+\Lambda_n}) .$$

In order to check that $\mu_1([\sigma_g]_{g\in\Lambda_n})$ is well defined, let $g'' \in \mathbb{Z}^d_+$ be such that $g'' + \Lambda_n \subset \mathbb{Z}^d_+$. There is g''' that dominates both g' and g'' coordinate-wise, i.e., there are $h', h'' \in \mathbb{Z}^d_+$ such that $g''' = g' + h'$ and $g''' = g'' + h''$. Then

$$\mu_2([\sigma_g]_{g\in g'''+\Lambda_n}) = \mu_2(T_2^{-h'}([\sigma_g]_{g\in g'+\Lambda_n})) = \mu_2([\sigma_g]_{g\in g'+\Lambda_n})$$

and the same holds for g'' instead of g'. Therefore μ_1 is well defined on cylinder sets. In the same way the finite additivity of μ_1 on cylinder sets is reduced to that of μ_2, and hence μ_1 extends uniquely to a measure on Ω_1, see A.4.8 of the appendix. Let $g \in \mathbb{Z}^d$. That $\mu_1(T^{-g}C) = \mu_1(C)$ for all cylinder sets in Ω_1 is checked in just the same way, and as $\mu_1 \circ T^{-g}$ and μ_1 are both probability measures, they must coincide, see A.4.7. It follows that $(\Omega_1, \mathcal{B}_1, \mu_1, T_1)$ is an extension of $(\Omega_2, \mathcal{B}_2, \mu_2, T_2)$. Obviously $\mathcal{B}_1 = \sigma(\bigcup_{g\in\mathbb{Z}^d} T^{-g}\Phi^{-1}\mathcal{B}_2)$, so it is in fact a natural extension. ◇

2.5.2 Exercise Give the details of the above proof that to each $\mathcal{T}_2$-invariant measure μ_2 there corresponds a unique $\mathcal{T}_1$-invariant measure μ_1 such that $\mu_2 = \mu_1 \circ \Phi^{-1}$. ◇

2.5.3 Exercise Let $(X_1, \mathcal{B}_1, \mu_1, \mathcal{T}_1)$ be the baker's map from Exercise 1.4.8 and denote by $(X_2, \mathcal{B}_2, \mu_2)$ the unit interval with its Borel σ-algebra and Lebesgue measure. Define $T_2 : X_2 \to X_2, T(x) = 2x \bmod 1$. Show that $(X_1, \mathcal{B}_1, \mu_1, \mathcal{T}_1)$ is a natural extension of $(X_2, \mathcal{B}_2, \mu_2, T_2)$. ◇

2.5.4 Exercise Let $(X_1, \mathcal{B}_1, \mu_1, \mathcal{T}_1)$ be an extension of $(X_2, \mathcal{B}_2, \mu_2, \mathcal{T}_2)$.

a) Prove that $(X_2, \mathcal{B}_2, \mu_2, \mathcal{T}_2)$ is ergodic (weakly mixing, mixing) whenever the system $(X_1, \mathcal{B}_1, \mu_1, \mathcal{T}_1)$ is ergodic (weakly mixing, mixing).
b) Suppose $(X_1, \mathcal{B}_1, \mu_1, \mathcal{T}_1)$ is a natural extension of the ergodic (weakly mixing, mixing) system $(X_2, \mathcal{B}_2, \mu_2, \mathcal{T}_2)$. Prove that $(X_1, \mathcal{B}_1, \mu_1, \mathcal{T}_1)$ is ergodic (weakly mixing, mixing). ◇

The essential conclusion of this discussion is that there is a one-to-one correspondence between invariant measures on the "one-sided" and on the "two-sided" shift spaces. In particular, each shift-invariant measure on the one-sided space extends to a unique shift-invariant measure on the two-sided space, thereby preserving ergodicity, weak mixing and mixing.

Chapter 3

Entropy

In this chapter we extend the notion of entropy per lattice site from shift-invariant measures on finite lattices (see Example 1.2.1) to shift-invariant measures on infinite lattices and, more generally, to measures which are invariant under some (semi)group $\mathcal{T}$ of transformations. The presentation of the material is influenced by many textbooks, among them [38], [42], [43], [46], [60].

We start with a section where the notion of entropy of a probability vector from Section 1.1 is generalized to a concept of (conditional) entropy of partitions on general probability spaces. Dynamics are added to this in Section 3.2, where the entropy of a m.p.d.s. $(X, \mathcal{B}, \mathcal{T}, \mu)$ is defined and its basic properties are explored. In the last section we slightly change our point of view: since a given action $\mathcal{T}$ on a measurable space $(X, \mathcal{B})$ may have many invariant measures, we look at the entropy of a dynamical system as a function of the measure.

3.1 Information and entropy of partitions

Let $(X, \mathcal{B}, \mu)$ be a probability space. A collection $\alpha = \{A_i : i \in I\}$ of measurable subsets of X is a *μ-partition* of X if $\mu(A_i \cap A_j) = 0$ for $i \neq j$, $\mu(X \setminus \bigcup_{i \in I} A_i) = 0$ and $\mu(A_i) > 0$ for all $i \in I$. Because of the last requirement, α is at most countable.

In later sections we also need partitions in a more narrow sense. We say that a finite or countable collection $\alpha = \{A_i : i \in I\}$ of nonempty $\mathcal{B}$-measurable sets is a *partition* of X, if the A_i are pairwise disjoint and if $\bigcup_{i \in I} A_i = X$.

In a dynamical context partitions may be understood as devices to make coarse-grained observations on points x from the state space X. Instead of observing the "coordinates" of a point precisely, only the index $i = i(x) \in I$ with "$x \in A_i$" is registered. If x is a random point with respect to μ, the knowledge of $i(x)$ contains the more information the better x is localizable by it, i.e., the smaller $\mu(A_{i(x)})$ is.

Let β be another μ-partition of X. The *common refinement* of α and β is the

μ-partition

$$\alpha \vee \beta := \{A \cap B : A \in \alpha, B \in \beta, \mu(A \cap B) > 0\} .$$

$\alpha \vee \beta$ is the partition that represents the combined outcome of (simultaneous or successive) observation by α and β. If α and β are independent (i.e., $\mu(A \cap B) = \mu(A)\mu(B)\ \forall A \in \alpha, B \in \beta$), any reasonable quantification $I_\alpha(x)$ of the concept of information of a μ-partition α should satisfy $I_{\alpha\vee\beta}(x) = I_\alpha(x) + I_\beta(x)$ for all $x \in X$, cf. Remark 1.1.1. As the only continuous (in fact, the only measurable) functions $I : [0,1] \to \mathbb{R}$ which satisfy the functional equation $I(xy) = I(x)+I(y)$ are the logarithms, one is naturally led to the following definition:

3.1.1 Definition *The information of the μ-partition α is*

$$\begin{aligned} I_\alpha(x) &:= -\sum_{A\in\alpha} \log \mu(A) \cdot 1_A(x) \\ &= -\log \mu(A) \quad \text{for } \mu\text{-a.e. } x \in A \in \alpha. \end{aligned}$$

For mathematical convenience we choose the natural logarithm for log. *The average information*

$$H(\alpha) := \int I_\alpha \, d\mu = -\sum_{A\in\alpha} \mu(A) \log \mu(A) = \sum_{A\in\alpha} \varphi(\mu(A))$$

(where $\varphi(t) = -t \log t$ as in Chapter 1) is called the entropy of α.

For finite μ-partitions, $H(\alpha)$ thus coincides with the entropy of the probability vector $(\mu(A))_{A\in\alpha}$ defined in Section 1.1 and

$$H(\alpha) \le \log \operatorname{card}(\alpha) , \tag{3.1}$$

see Remark 1.1.1. For infinite μ-partitions the entropy $H(\alpha)$ may be infinite. This is the case if, e.g., $\alpha = (A_n : n > 1)$ with $\mu(A_n) = Z^{-1}n^{-1}(\log n)^{-2}$ where Z is a suitable norming constant.

More refined versions of information and entropy are needed, if one wants to describe the gain of information of an observer who already has some knowledge about the system and observes it now through a partition α. The knowledge he starts with can be represented by a sub-σ-algebra $\mathcal{F}$ of $\mathcal{B}$ in the sense that for each $x \in X$ and each set $F \in \mathcal{F}$ the observer knows whether $x \in F$. Before we make this idea more precise, we note the following lemma:

3.1.2 Lemma *Let α be a μ-partition and let $\mathcal{F}$ be a sub-σ-algebra of $\mathcal{B}$. Then $\mu(A \mid \mathcal{F}) > 0$ μ-a.e. on A for each $A \in \alpha$.*

Proof: For $A \in \alpha$ fix an arbitrary version of $\mu(A \mid \mathcal{F})$ and let $F_A = \{x \in X : \mu(A \mid \mathcal{F})(x) = 0\}$. Then $F_A \in \mathcal{F}$ and

$$\mu(A \cap F_A) = \int_{F_A} 1_A \, d\mu = \int_{F_A} \mu(A \mid \mathcal{F}) \, d\mu = 0$$

so that $\mu(A \mid \mathcal{F}) > 0$ μ-a.e. on A. □

This lemma allows us to make the following definition:

3.1.3 Definition *Let $\mathcal{F}$ be a sub-σ-algebra of $\mathcal{B}$ and α a μ-partition.*

$$I_{\alpha|\mathcal{F}}(x) := -\sum_{A \in \alpha} \log \mu(A \mid \mathcal{F})(x) \cdot 1_A(x)$$

is the conditional information of α given $\mathcal{F}$.

$$H(\alpha \mid \mathcal{F}) := \int I_{\alpha|\mathcal{F}} \, d\mu = \sum_{A \in \alpha} \int \varphi \circ \mu(A \mid \mathcal{F}) \, d\mu$$

is the conditional entropy of α given $\mathcal{F}$.

The following assertions are easily verified:

- ▷ $I_{\alpha|\mathcal{F}} \geq 0$ μ-a.s. and hence also $H(\alpha \mid \mathcal{F}) \geq 0$.
- ▷ If $\mathcal{F}$ is *trivial* mod μ, i.e., if $\mu(F) \in \{0, 1\}$ for all $F \in \mathcal{F}$, then $I_{\alpha|\mathcal{F}} = I_\alpha$ μ-a.s. and $H(\alpha \mid \mathcal{F}) = H(\alpha)$.
- ▷ If $\mathcal{F} = \mathcal{B}$ mod μ, i.e., if $\mu(A \mid \mathcal{F}) = 1_A \in \{0, 1\}$ μ-a.s. for all $A \in \mathcal{B}$, then $I_{\alpha|\mathcal{F}} = 0$ μ-a.s. and hence also $H(\alpha \mid \mathcal{F}) = 0$.

In order to simplify the notation we write henceforth $\alpha \mid \beta$ instead of $\alpha \mid \sigma(\beta)$ where $\sigma(\beta)$ denotes the σ-algebra generated by the elements of β, and also $\alpha \mid \beta \vee \mathcal{F}$ instead of $\alpha \mid (\sigma(\beta) \vee \mathcal{F})$ where, as usual, $\mathcal{F} \vee \mathcal{G}$ denotes the smallest σ-algebra containing both $\mathcal{F}$ and $\mathcal{G}$. As $\sigma(\alpha \vee \beta) = \sigma(\alpha) \vee \sigma(\beta)$, this will cause no confusion. Observe also that since β is an at most countable partition, $\sigma(\beta)$ consists of all finite or countable unions of elements of β.

3.1.4 Remark With these notational conventions a conditional probability $\mu(A \mid \beta)$ can be written as $\mu(A \mid \beta) = \sum_{B \in \beta} \mu(A \mid B) \cdot 1_B$ where $\mu(A \mid B) = \frac{\mu(A \cap B)}{\mu(B)}$ as usual. Hence

$$\begin{aligned} H(\alpha \mid \beta) &= \sum_{A \in \alpha} \int \sum_{B \in \beta} \varphi(\mu(A \mid B)) \cdot 1_B \, d\mu = \sum_{A \in \alpha, B \in \beta} \varphi(\mu(A \mid B)) \cdot \mu(B) \\ &= -\sum_{A \in \alpha, B \in \beta} \mu(A \cap B) \log \mu(A \mid B) \,. \end{aligned}$$

A nice interpretation of this identity is $H(\alpha \mid \beta) = \sum_{B\in\beta} \mu(B) \cdot H(\alpha \mid B)$ where $H(\alpha \mid B)$ denotes the entropy of α w.r.t. the normalized restriction of μ to B. $\diamond$

For computations involving conditional information and conditional entropy the following theorem is nearly indispensable:

3.1.5 Theorem (Addition rule for information)
Let α and β be μ-partitions of X and let $\mathcal{F}$ be a sub-σ-algebra of $\mathcal{B}$. Then

$$I_{\alpha\vee\beta|\mathcal{F}} = I_{\alpha|\mathcal{F}} + I_{\beta|\alpha\vee\mathcal{F}} \quad \mu\text{-a.s.}$$

Proof: We start by proving that for each $B \in \beta$

$$\mu(B \mid \alpha \vee \mathcal{F}) = \sum_{A\in\alpha} \frac{\mu(A \cap B \mid \mathcal{F})}{\mu(A \mid \mathcal{F})} \cdot 1_A \quad \mu\text{-a.s.} \tag{3.2}$$

To this end observe that

1. $\mu(A \mid \mathcal{F}) > 0$ μ-a.s. on A for each $A \in \alpha$ by Lemma 3.1.2,
2. the right hand side of equation (3.2) is $\alpha \vee \mathcal{F}$-measurable, and
3. for each $A' \in \alpha$ and $F \in \mathcal{F}$

$$\begin{aligned}
\int_{A'\cap F} \left(\sum_{A\in\alpha} \frac{\mu(A \cap B \mid \mathcal{F})}{\mu(A \mid \mathcal{F})} \cdot 1_A \right) d\mu &= \int_F 1_{A'} \frac{\mu(A' \cap B \mid \mathcal{F})}{\mu(A' \mid \mathcal{F})} \, d\mu \\
&= \int_F E[1_{A'} \mid \mathcal{F}] \cdot \frac{\mu(A' \cap B \mid \mathcal{F})}{\mu(A' \mid \mathcal{F})} \, d\mu = \int_F \mu(B \cap A' \mid \mathcal{F}) \, d\mu \\
&= \mu(B \cap A' \cap F) = \int_{A'\cap F} 1_B \, d\mu \, .
\end{aligned}$$

Taking logarithms on both sides of (3.2) yields

$$\log \mu(B \mid \alpha \vee \mathcal{F}) = \sum_{A\in\alpha} \big(\log \mu(A \cap B \mid \mathcal{F}) - \log \mu(A \mid \mathcal{F})\big) \cdot 1_A \quad \mu\text{-a.s.}$$

and hence

$$\begin{aligned}
& I_{\beta|\alpha\vee\mathcal{F}} \\
&= -\sum_{B\in\beta} \log \mu(B \mid \alpha \vee \mathcal{F}) \cdot 1_B \\
&= -\sum_{A\in\alpha} \sum_{B\in\beta} \log \mu(A \cap B \mid \mathcal{F}) \cdot 1_{A\cap B} + \sum_{B\in\beta} \left(\sum_{A\in\alpha} \log \mu(A \mid \mathcal{F}) \cdot 1_A \right) \cdot 1_B \\
&= I_{\alpha\vee\beta|\mathcal{F}} - I_{\alpha|\mathcal{F}} \quad \mu\text{-a.s.}
\end{aligned}$$

$\square$

3.1.6 Corollary *a)* $I_{\alpha\vee\beta} = I_\alpha + I_{\beta|\alpha}$ μ-a.s.
b) $H(\alpha \vee \beta \mid \mathcal{F}) = H(\alpha \mid \mathcal{F}) + H(\beta \mid \alpha \vee \mathcal{F})$.
c) $H(\alpha \vee \beta) = H(\alpha) + H(\beta \mid \alpha)$.
d) $H(\alpha) \leq H(\beta) + H(\alpha \mid \beta)$.

Proof: b) follows from Theorem 3.1.5 by integration, a) and c) are special cases for trivial $\mathcal{F}$, and d) follows from c) because $H(\alpha) = H(\alpha \vee \beta) - H(\beta \mid \alpha) \leq H(\alpha \vee \beta) = H(\beta) + H(\alpha \mid \beta)$. □

A μ-partition β is *finer* than α ($\beta \succeq \alpha$), if each $A \in \alpha$ is a union of elements of β (mod μ). If $\beta \succeq \alpha$ and $\alpha \succeq \beta$, we write $\alpha \approx \beta$. In this case α and β coincide mod μ.

3.1.7 Lemma $\beta \succeq \alpha \iff \alpha \vee \beta \approx \beta \iff H(\alpha \mid \beta) = 0$.

Proof: The first equivalence is true by definition. For the second one observe that $H(\alpha \mid \beta) = \sum_{A\in\alpha, B\in\beta} \varphi(\mu(A \mid B)) \cdot \mu(B)$ and $\varphi(\mu(A \mid B)) \geq 0$ with equality if and only if $\mu(A \mid B) \in \{0, 1\}$ so that

$$H(\alpha \mid \beta) = 0 \iff \forall A \in \alpha \forall B \in \beta : \mu(A \mid B) \in \{0,1\} \iff \beta \succeq \alpha .$$

□

Observe also that $I_{\alpha|\mathcal{F}} = I_{\beta|\mathcal{F}}$ μ-a.s. and hence $H(\alpha \mid \mathcal{F}) = H(\beta \mid \mathcal{F})$ if $\alpha \approx \beta$. Therefore we need not distinguish between μ-partitions and their "$\approx$"-equivalence classes as long as only one measure μ and countably many partitions are involved in our arguments.

3.1.8 Theorem (Monotonicity of conditional entropy)
Let α and β be μ-partitions of X and $\mathcal{F}, \mathcal{F}_1, \mathcal{F}_2$ sub-σ-algebras of $\mathcal{B}$.
a) If $\mathcal{F}_1 \supseteq \mathcal{F}_2$, then $H(\alpha \mid \mathcal{F}_1) \leq H(\alpha \mid \mathcal{F}_2)$ and in particular $H(\alpha \mid \mathcal{F}_1) \leq H(\alpha)$.
b) If $\beta \succeq \alpha$, then $H(\beta \mid \mathcal{F}) \geq H(\alpha \mid \mathcal{F})$ and in particular $H(\beta) \geq H(\alpha)$.
c) $H(\alpha \vee \beta \mid \mathcal{F}) \leq H(\alpha \mid \mathcal{F}) + H(\beta \mid \mathcal{F})$, in particular $H(\alpha \vee \beta) \leq H(\alpha) + H(\beta)$.

Proof:
a) Recall that $\varphi(t) = -t \log t$. By the conditional version of Jensen's inequality (see A.4.35)

$$E[\varphi \circ \mu(A \mid \mathcal{F}_1) \mid \mathcal{F}_2] \leq \varphi \circ E[\mu(A \mid \mathcal{F}_1) \mid \mathcal{F}_2] = \varphi \circ \mu(A \mid \mathcal{F}_2) ,$$

where $\mathcal{F}_1 \supseteq \mathcal{F}_2$ is used for the equality. Integration with respect to μ and summation over $A \in \alpha$ yield

$$H(\alpha \mid \mathcal{F}_1) = \sum_{A\in\alpha} \int \varphi \circ \mu(A \mid \mathcal{F}_1)\, d\mu \leq \sum_{A\in\alpha} \int \varphi \circ \mu(A \mid \mathcal{F}_2)\, d\mu = H(\alpha \mid \mathcal{F}_2) .$$

b) $H(\alpha \mid \mathcal{F}) \le H(\alpha \mid \mathcal{F}) + H(\beta \mid \alpha \vee \mathcal{F}) = H(\alpha \vee \beta \mid \mathcal{F}) = H(\beta \mid \mathcal{F})$ by Corollary 3.1.6-b and Lemma 3.1.7.

c) $H(\alpha \vee \beta \mid \mathcal{F}) = H(\alpha \mid \mathcal{F}) + H(\beta \mid \alpha \vee \mathcal{F}) \le H(\alpha \mid \mathcal{F}) + H(\beta \mid \mathcal{F})$ by Corollary 3.1.6-b and assertion a) of this theorem. □

3.1.9 Exercise Prove the equivalence of the following statements concerning μ-partitions α and β of $(X, \mathcal{B}, \mu)$:

1. α and β are independent.
2. $H(\alpha \vee \beta) = H(\alpha) + H(\beta)$.
3. $H(\alpha \mid \beta) = H(\alpha)$. ◇

3.1.10 Theorem (Continuity of conditional information and entropy) *Let $\mathcal{F}_1 \subseteq \mathcal{F}_2 \subseteq \mathcal{F}_3 \subseteq \dots$ be sub-σ-algebras of $\mathcal{B}$, and denote by $\mathcal{F}$ the σ-algebra generated by all $\mathcal{F}_n$. For any μ-partition α of finite entropy*

$$\lim_{n\to\infty} I_{\alpha|\mathcal{F}_n} = I_{\alpha|\mathcal{F}} \ \mu\text{-a.s. and in } L^1_\mu, \quad \text{and} \quad \sup_{n>0} I_{\alpha|\mathcal{F}_n} \in L^1_\mu .$$

In particular, $\lim_{n\to\infty} H(\alpha \mid \mathcal{F}_n) = H(\alpha \mid \mathcal{F})$.

Proof: Let $A \in \alpha$. A corollary of the martingale convergence theorem (A.4.38) yields $\lim_{n\to\infty} \mu(A \mid \mathcal{F}_n) = \mu(A \mid \mathcal{F})$ μ-a.s. Therefore

$$\lim_{n\to\infty} I_{\alpha|\mathcal{F}_n} = -\sum_{A\in\alpha} 1_A \cdot \lim_{n\to\infty} \log \mu(A \mid \mathcal{F}_n) = I_{\alpha|\mathcal{F}} \quad \mu\text{-a.s.},$$

and, because of Lebesgue's dominated convergence theorem (A.4.20), the L^1_μ-convergence follows once we have shown that $f^* := \sup_{n>0} I_{\alpha|\mathcal{F}_n}$ is μ-integrable. To this end we estimate $\mu(A \cap \{f^* > t\})$ for arbitrary $t > 0$ and $A \in \alpha$. Let $f_{n,A} := -\log \mu(A \mid \mathcal{F}_n)$ and

$$C_{n,A}(t) := \{x \in X : \ f_{1,A}(x) \le t, \dots, f_{n-1,A}(x) \le t, f_{n,A}(x) > t\} .$$

As $C_{n,A}(t)$ is $\mathcal{F}_n$-measurable, we have

$$\mu(A \cap C_{n,A}(t)) = \int_{C_{n,A}(t)} \mu(A \mid \mathcal{F}_n)\, d\mu = \int_{C_{n,A}(t)} e^{-f_{n,A}}\, d\mu \le e^{-t} \cdot \mu(C_{n,A}(t))$$

and therefore

$$\mu(A \cap \{f^* > t\}) = \sum_{n=1}^{\infty} \mu(A \cap C_{n,A}(t)) \le e^{-t} \cdot \sum_{n=1}^{\infty} \mu(C_{n,A}(t)) \le e^{-t}$$

so that

$$
\begin{aligned}
\int f^* \, d\mu &= \int_0^\infty \mu\{f^* > t\} \, dt \\
&= \sum_{A \in \alpha} \int_0^\infty \mu(A \cap \{f^* > t\}) \, dt \\
&\leq \sum_{A \in \alpha} \int_0^\infty \min\{e^{-t}, \mu(A)\} \, dt \\
&= \sum_{A \in \alpha} \left(\int_0^{-\log \mu(A)} \mu(A) \, dt + \int_{-\log \mu(A)}^\infty e^{-t} \, dt \right) \\
&= H(\alpha) + \sum_{A \in \alpha} \mu(A) = H(\alpha) + 1 \\
&< \infty .
\end{aligned}
$$

□

3.2 Entropy of dynamical systems

Now, as already in Chapter 2, $(X, \mathcal{B}, \mu, \mathcal{T})$ is again a dynamical system, i.e., $\mathcal{T}$ denotes an action of $G = \mathbb{Z}^d$ or $G = \mathbb{Z}_+^d$ on X which leaves the measure μ invariant. We use the following notations:

- ▷ For a μ-partition α of X and $g \in G$ denote by $T^{-g}\alpha := \{T^{-g}A : A \in \alpha\}$ the μ-partition of X into T^g-preimages of the sets $A \in \alpha$. If α is a partition in the strict sense, then so is $T^{-g}\alpha$.

- ▷ For a finite subset $\Lambda \subseteq G$ let $\alpha^\Lambda := \bigvee_{g \in \Lambda} T^{-g}\alpha$ be the common refinement of the μ-partitions $T^{-g}\alpha$, $g \in \Lambda$. If α is a partition in the strict sense, then so is α^Λ. With these notations it is equivalent to screen a random point $x \in X$ through the partition α^Λ or to screen all points $T^g x$ $(g \in \Lambda)$ through α. As before let $\Lambda_n = \{g \in G : |g_i| < n \, \forall i\}$ and $\lambda_n = |\Lambda_n|$.

- ▷ For infinite $\tilde{G} \subseteq G$ denote by $\alpha^{\tilde{G}}$ the *σ-algebra* generated by all $T^{-g}\alpha$, $g \in \tilde{G}$.

- ▷ In the case $d = 1$ we also use the following notation: For $k < n$ let $\alpha_k^n := \alpha^{\{k,\dots,n-1\}}$, and by α_k^∞ we denote the σ-algebra generated by all α_k^n, $n > k$. By convention, $\alpha_k^k = \{X\}$ is the trivial partition.

In order to understand the development of the theory in this section it is helpful to keep in mind the following examples:

3.2.1 Example (Entropy in shift spaces) (Continuation of Example 2.2.10) Let Σ be a finite set, $\Omega = \Sigma^G$, $G = \mathbb{Z}^d$ or $G = \mathbb{Z}_+^d$, and let $T^g : \Omega \to \Omega$ denote the shift by g. For $\sigma \in \Sigma$ let $[\sigma]_0 := \{\omega \in \Omega : \omega_0 = \sigma\}$. (A comparison of this notation with the one introduced for cylinder sets in (2.12) shows that $[\sigma]_0$ is simply shorthand for $[\sigma_g]_{g\in\{0\}}$ with $\sigma_0 = \sigma$.) Then $\alpha = \{[\sigma]_0 : \sigma \in \Sigma\}$ is a finite partition of Ω and α^{Λ_n} coincides with the partition $\mathcal{C}_n$ into cylinder sets introduced in Example 2.2.10. Let μ be any $\mathcal{T}$-invariant measure on Ω. Then $I_{\alpha^{\Lambda_n}}(\omega)$ is the amount of information on the configuration ω contained in the coordinates ω_g, $g \in \Lambda_n$, and $H(\alpha^{\Lambda_n})$ is the average of this amount over all configurations. Similarly, $I_{\alpha^{\Lambda_{n+1}}|\alpha^{\Lambda_n}}(\omega)$ is the additional information gained when observing the ω_g, $g \in \Lambda_{n+1} \setminus \Lambda_n$, if the ω_g, $g \in \Lambda_n$, are already known. ◇

3.2.2 Example Let $\mathcal{T} = (T^k \mid k \in \mathbb{Z}_+)$. If $T : X \to X$ describes the time evolution of a dynamical system and if α is a finite partition, then $I_{\alpha^{\Lambda_n}}(\omega)$ is the total amount of information on ω gained by observing the system through α at times $0, \ldots, n-1$. In this context $I_{\alpha^{\Lambda_{n+1}}|\alpha^{\Lambda_n}}(\omega)$ is the additional information gained at time n when all observations up to time $n-1$ are known. ◇

3.2.3 Lemma *For each $g \in G$, each μ-partition α and each sub-σ-algebra $\mathcal{F} \subset \mathcal{B}$ we have*

$$I_{T^{-g}\alpha|T^{-g}\mathcal{F}} = I_{\alpha|\mathcal{F}} \circ T^g \quad \text{and} \quad H(T^{-g}\alpha \mid T^{-g}\mathcal{F}) = H(\alpha \mid \mathcal{F}) .$$

In particular, $I_{T^{-g}\alpha} = I_\alpha \circ T^g$ and $H(T^{-g}\alpha) = H(\alpha)$.

Proof: As for each $F \in \mathcal{F}$ and $A \in \alpha$

$$\begin{aligned}\int_{T^{-g}F} \mu(A \mid \mathcal{F}) \circ T^g \, d\mu &= \int_F \mu(A \mid \mathcal{F})\, d\mu = \mu(A \cap F) = \mu(T^{-g}A \cap T^{-g}F) \\ &= \int_{T^{-g}F} 1_{T^{-g}A} \, d\mu ,\end{aligned}$$

we have $\mu(T^{-g}A \mid T^{-g}\mathcal{F}) = \mu(A \mid \mathcal{F}) \circ T^g$ μ-a.e. Therefore

$$\begin{aligned}I_{T^{-g}\alpha|T^{-g}\mathcal{F}} &= -\sum_{A\in\alpha} \log \mu(T^{-g}A \mid T^{-g}\mathcal{F}) \cdot 1_{T^{-g}A} \\ &= -\sum_{A\in\alpha} (\log \mu(A \mid \mathcal{F}) \cdot 1_A) \circ T^g = I_{\alpha|\mathcal{F}} \circ T^g .\end{aligned}$$

Observing that $T^g\mu = \mu$, the identity for the entropy follows by integrating the identity for the information. □

The behaviour of $H(\alpha^{\Lambda_n})$ as $n \to \infty$ is described by the next theorem:

3.2.4 Theorem (Dynamical entropy relative to a partition)
Let $(X, \mathcal{B}, \mu, \mathcal{T})$ be a m.p.d.s. and let α be a μ-partition of X with $H(\alpha) < \infty$. Then

$$h(\alpha, \mathcal{T}) := \inf_{n>0} \frac{1}{\lambda_n} H(\alpha^{\Lambda_n}) = \lim_{n\to\infty} \frac{1}{\lambda_n} H(\alpha^{\Lambda_n}) \tag{3.3}$$

is the dynamical entropy of the system $(X, \mathcal{B}, \mu, \mathcal{T})$ relative to the partition α. If $H(\alpha) = \infty$, then $H(\alpha^{\Lambda_n}) = \infty$ for all n and hence $h(\alpha, \mathcal{T}) = \infty$.

Proof: Let l_n be the side length of the "cube" Λ_n $(n > 0)$. Fix some $m > 0$ and let $V_n = \Lambda_n \cap (l_m\mathbb{Z})^d$. Then $\Lambda_n \subseteq \tilde{\Lambda}_n := \bigcup_{g \in V_n} (\Lambda_m + g)$ and $|\tilde{\Lambda}_n| = |V_n| \cdot \lambda_m \leq \lambda_{n+m}$. (The situation is sketched in Figure 3.1 for $G = \mathbb{Z}_+^2$, $m = 3$ and $n = 8$.) It follows from Theorem 3.1.8 and Lemma 3.2.3 that

$$\begin{aligned} H(\alpha^{\Lambda_n}) &\leq H(\alpha^{\tilde{\Lambda}_n}) \leq \sum_{g \in V_n} H(T^{-g} \alpha^{\Lambda_m}) \\ &= |V_n| \cdot H(\alpha^{\Lambda_m}) \\ &\leq \frac{\lambda_{n+m}}{\lambda_m} \cdot H(\alpha^{\Lambda_m}) . \end{aligned}$$

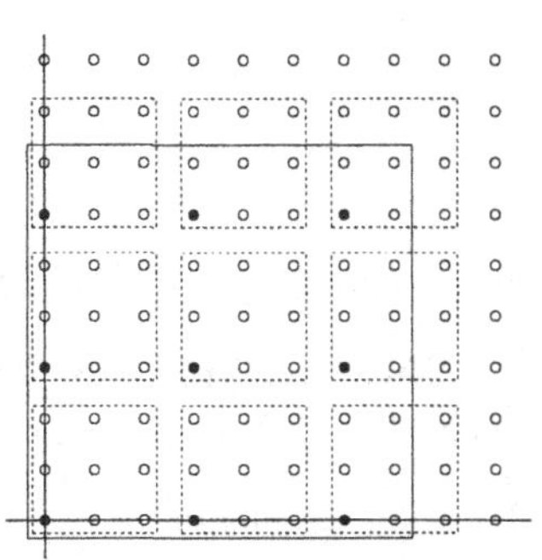

Figure 3.1: Sketch of the sets V_n (black dots), $\tilde{\Lambda}_n$ (union of the dashed boxes) and Λ_n (solid box).

Therefore

$$\limsup_{n\to\infty} \frac{1}{\lambda_n} \cdot H(\alpha^{\Lambda_n}) \leq \limsup_{n\to\infty} \underbrace{\frac{\lambda_{n+m}}{\lambda_n}}_{\to 1} \cdot \frac{1}{\lambda_m} H(\alpha^{\Lambda_m}) = \frac{1}{\lambda_m} \cdot H(\alpha^{\Lambda_m}) .$$

As this holds for each fixed m, we conclude that

$$\limsup_{n\to\infty} \frac{1}{\lambda_n} \cdot H(\alpha^{\Lambda_n}) \leq \inf_{m>0} \frac{1}{\lambda_m} \cdot H(\alpha^{\Lambda_m}) \leq \liminf_{m\to\infty} \frac{1}{\lambda_m} \cdot H(\alpha^{\Lambda_m}) ,$$

and the convergence as well as the identity in (3.3) follows immediately. □

Any $\mathbb{Z}^d$-action $\mathcal{T}$ can be restricted to a $\mathbb{Z}_+^d$-action $\mathcal{T}^+$. Obviously any $\mathcal{T}$-invariant measure μ is also $\mathcal{T}^+$-invariant, and one may ask whether $h(\alpha, \mathcal{T}) = h(\alpha, \mathcal{T}^+)$.

3.2.5 Corollary *In the situation of the previous theorem suppose that $G = \mathbb{Z}^d$. Then*

$$h(\alpha, \mathcal{T}) = h(\alpha, \mathcal{T}^+) .$$

Proof: Denote $\Lambda_n^+ := \{g \in \mathbb{Z}^d : 0 \leq g_i < n \ \forall i = 1, \ldots, d\}$ and let $1 = (1, \ldots, 1) \in \mathbb{Z}^d$. Then

$$H(\alpha^{\Lambda_n}) = H(T^{-(n-1)1} \alpha^{\Lambda_{2n-1}^+}) = H(\alpha^{\Lambda_{2n-1}^+}) ,$$

and the assertion follows from the observation $|\Lambda_n| = |\Lambda_{2n-1}^+|$. □

3.2.6 Remark If $d = 1$, i.e., if $\mathcal{T}$ is generated by a single transformation T, we define

$$h(\alpha, T) := \lim_{n\to\infty} \frac{1}{n} H(\alpha_0^n) .$$

In the case $G = \mathbb{Z}_+$ we have $h(\alpha, T) = h(\alpha, \mathcal{T})$ by definition. If $G = \mathbb{Z}$, i.e., if T is invertible, the previous corollary implies

$$h(\alpha, T) = h(\alpha, \mathcal{T}) = h(\alpha, T^{-1}) .$$

◇

The asymptotic properties of $I_{\alpha^{\Lambda_n}}$ are clarified by

3.2.7 Theorem (Shannon–McMillan–Breiman)
Let $(X, \mathcal{B}, \mu, \mathcal{T})$ be a m.p.d.s. and let α be a μ-partition of X with $H(\alpha) < \infty$.

a) *For all dimensions d*

$$\lim_{n\to\infty} \frac{1}{|\Lambda_n^+|} I_{\alpha^{\Lambda_n^+}} = E[I_{\alpha|\alpha^{G^-}} \mid \mathcal{I}_\mu(\mathcal{T})] \quad \text{in } L_\mu^1$$

where G^- denotes the set of all $g \in \mathbb{Z}^d$ that precede 0 in the lexicographic ordering of G. In particular, $h(\alpha, \mathcal{T}) = H(\alpha \mid \alpha^{G^-})$. If α is finite, then this convergence is also μ-a.s. (If $G = \mathbb{Z}_+^d$, then $E[I_{\alpha|\alpha^{G^-}} \mid \mathcal{I}_\mu(\mathcal{T})]$ must be taken in the natural extension of $(X, \mathcal{B}, \mu, \mathcal{T})$, see [34, p. 283].)

b) *If $d = 1$ even more holds:*

$$\lim_{n\to\infty} \frac{1}{n} I_{\alpha_0^n} = E[I_{\alpha|\alpha_1^\infty} \mid \mathcal{I}_\mu(T)] \quad \text{and} \quad \lim_{n\to\infty} I_{\alpha|\alpha_1^n} = I_{\alpha|\alpha_1^\infty}$$

both μ-a.s. and in L_μ^1. The sequence $(\frac{1}{n} H(\alpha_0^n))_{n>0}$ decreases monotonically to $h(\alpha, T)$ and

$$h(\alpha, T) = \lim_{n\to\infty} H(\alpha \mid \alpha_1^n) = H(\alpha \mid \alpha_1^\infty) .$$

(These two equalities are obtained by integrating the corresponding identities for the information.)

3.2.8 Corollary *If $(X, \mathcal{B}, \mu, \mathcal{T})$ is ergodic and if $H(\alpha) < \infty$, then*

$$h(\alpha, \mathcal{T}) = \lim_{n\to\infty} \frac{1}{|\Lambda_n^+|} I_{\alpha^{\Lambda_n^+}} \quad \text{in } L_\mu^1$$

and

$$h(\alpha, T) = \lim_{n\to\infty} \frac{1}{n} I_{\alpha_0^n} \quad \mu\text{-a.s. and in } L_\mu^1 \text{ (in the case } d = 1).$$

3.2.9 Remark The expression $H(\alpha \mid \alpha^{G^-})$ for $h(\alpha, T)$ has the following interpretation which can best be visualized for $d = 2$:
If the cube Λ_n^+ is built up step by step in lexicographic order, then at most steps the "local" picture at the boundary of the "growing" Λ_n is the same as that of G^- at position 0, see Figure 3.2. Hence $H(\alpha \mid \alpha^{G^-})$ is the approximate value of the information gain at most steps of the enlargement procedure. For $d = 1$ this idea is made precise in the subsequent proof. ◇

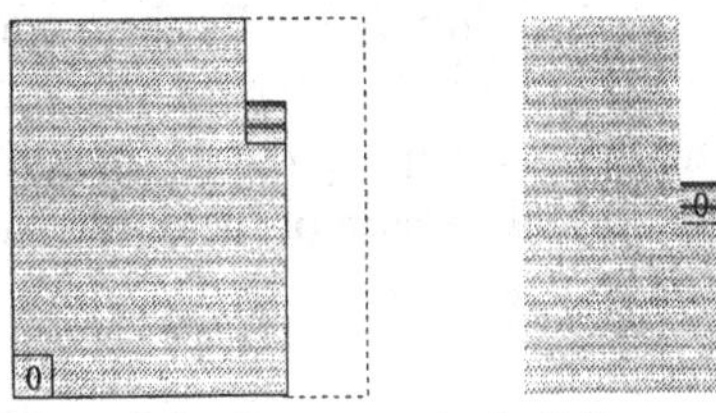

Figure 3.2: One step in building up Λ_n (l.h.s.) and the set $G^- \subset \mathbb{Z}^2$ (r.h.s.).

Proof of the theorem:

a) For a proof and further information in case $d > 1$ we refer to [34, Theorems 9.2.4 and 9.2.5].

b) In view of Corollary 3.1.6-a and Lemma 3.2.3 we have

$$I_{\alpha_0^n} = I_{\alpha \vee T^{-1}\alpha_0^{n-1}} = I_{T^{-1}\alpha_0^{n-1}} + I_{\alpha | T^{-1}\alpha_0^{n-1}} = I_{\alpha_0^{n-1}} \circ T + I_{\alpha|\alpha_1^n} \,.$$

Applying this identity repeatedly we obtain

$$I_{\alpha_0^n} = I_{\alpha|\alpha_1^n} + I_{\alpha|\alpha_1^{n-1}} \circ T + \cdots + I_{\alpha|\alpha_1^2} \circ T^{n-2} + I_{\alpha|\alpha_1^1} \circ T^{n-1} \,. \tag{3.4}$$

(Observe that $\alpha_0^1 = \alpha$ and $\alpha_1^1 = (X)$.) Hence

$$\begin{aligned} \frac{1}{n} I_{\alpha_0^n} &= \frac{1}{n}\sum_{k=1}^{n} I_{\alpha|\alpha_1^k} \circ T^{n-k} = \frac{1}{n}\sum_{k=0}^{n-1} I_{\alpha|\alpha_1^{n-k}} \circ T^k \\ &= \underbrace{\frac{1}{n}\sum_{k=0}^{n-1} I_{\alpha|\alpha_1^\infty} \circ T^k}_{=:R_n} + \underbrace{\frac{1}{n}\sum_{k=0}^{n-1} (I_{\alpha|\alpha_1^{n-k}} - I_{\alpha|\alpha_1^\infty}) \circ T^k}_{=:S_n} \,. \end{aligned}$$

As $E[I_{\alpha|\alpha_1^\infty}] = H(\alpha \mid \alpha_1^\infty) \leq H(\alpha) < \infty$, the average R_n converges in view of the ergodic theorem 2.1.5 almost surely and in L_μ^1 to $E[I_{\alpha|\alpha_1^\infty} \mid \mathcal{I}_\mu(T)]$. So it remains to prove that $S_n \to 0$ almost surely and in L_μ^1. For the L_μ^1-convergence observe that

$$\int |S_n| \, d\mu \leq \frac{1}{n}\sum_{k=0}^{n-1} \int |I_{\alpha|\alpha_1^{n-k}} - I_{\alpha|\alpha_1^\infty}| \, d\mu \to 0$$

by Theorem 3.1.10. The proof of the almost sure convergence needs a bit more thought:

Let $h_N := \sup_{n\ge N} |I_{\alpha|\alpha_1^n} - I_{\alpha|\alpha_1^\infty}|$. Because of Theorem 3.1.10,

$$\lim_{N\to\infty} h_N = 0 \ \mu\text{-a.s.} \quad \text{and} \quad h_N \le 2f^* \in L^1_\mu \text{ for all } N > 0,$$

where $f^* = \sup_{n>0} I_{\alpha|\alpha_1^n}$. Therefore $\lim_{N\to\infty} E[h_N \mid \mathcal{I}_\mu(T)] = 0$ μ-a.s. by the conditional version of the dominated convergence theorem (A.4.36). On the other hand,

$$\begin{aligned} \limsup_{n\to\infty} |S_n| &\le \limsup_{n\to\infty} \left(\frac{1}{n} \sum_{k=0}^{n-N} h_N \circ T^k + \frac{1}{n} \sum_{k=n-N+1}^{n-1} h_1 \circ T^k \right) \\ &= E[h_N \mid \mathcal{I}_\mu(T)] \quad \mu\text{-a.s.} \end{aligned}$$

for each $N > 0$ by another application of Birkhoff's ergodic theorem. Therefore $\lim_{n\to\infty} S_n = 0$ μ-a.s. follows for $N \to \infty$.

The monotonicity of the sequence $(\frac{1}{n} H(\alpha_0^n))_{n>0}$ can be seen as follows: integrating (3.4) one obtains

$$H(\alpha_0^n) = H(\alpha) + H(\alpha \mid \alpha_1^2) + \cdots + H(\alpha \mid \alpha_1^{n-1}) + H(\alpha \mid \alpha_1^n),$$

and as the sequence $(H(\alpha \mid \alpha_1^n))_{n>0}$ decreases monotonically (Theorem 3.1.8-a), $(\frac{1}{n} H(\alpha_0^n))_{n>0}$ is also decreasing. □

3.2.10 Exercise Let $(X, \mathcal{B}, \mu, \mathcal{T})$ be a m.p.d.s. and let $0 < r_1^{(i)} < r_2^{(i)} < r_3^{(i)} < \dots$ be d sequences of integers ($i = 1, \dots, d$) defining "parameter rectangles" $V_n := \{g \in G : |g_i| < r_n^{(i)} \ \forall i = 1, \dots, d\}$. Prove that

$$\lim_{n\to\infty} \frac{1}{|V_n|} H(\alpha^{V_n}) = h(\alpha, \mathcal{T}).$$

◇

In a final step we free the notion of entropy of a dynamical system from the reference to an arbitrary partition.

3.2.11 Definition *The entropy of the m.p.d.s.* $(X, \mathcal{B}, \mu, \mathcal{T})$ *is*

$$h_\mathcal{T}(\mu) := \sup\{h(\alpha, \mathcal{T}) : \alpha \in \mathcal{P}_0\},$$

where $\mathcal{P}_0$ *denotes the set of all finite partitions of* $(X, \mathcal{B})$. *Later we suppress notationally the dependence on* $\mathcal{T}$ *and write* $h(\mu)$ *instead of* $h_\mathcal{T}(\mu)$.

If $G \subseteq \mathbb{Z}$ and if $\mathcal{T}$ describes the time evolution of a system, the entropy $h_\mathcal{T}(\mu)$ can thus be interpreted as the maximal information per unit of time that an observer of the system can gain, if he looks at it through any finite partition.

3.2.12 Remarks

a) In order to distinguish the dynamical entropy $h_{\mathcal{T}}(\mu)$ from other entropies this quantity is often called *Kolmogorov–Sinai entropy* thus honouring two of the pioneers of entropy theory for dynamical systems.

b) In many textbooks the notation $h_\mu(\mathcal{T})$ is used instead of $h_{\mathcal{T}}(\mu)$, but because we are mostly dealing with a fixed action $\mathcal{T}$ and varying measure μ, we prefer $h_{\mathcal{T}}(\mu)$. If it is necessary to make the dependence of $H(\alpha)$ and $h(\alpha, \mathcal{T})$ on μ more explicit, we also write $H_\mu(\alpha)$ and $h(\mu, \alpha, \mathcal{T})$ respectively. ◇

The following theorem shows that observing a system through a countable partition of finite entropy does not give more information than looking at it through finite partitions.

3.2.13 Theorem (Entropy via countable partitions)
Let $(X, \mathcal{B}, \mu, \mathcal{T})$ be a m.p.d.s. Then

$$h_{\mathcal{T}}(\mu) = \sup\{h(\alpha, \mathcal{T}) : \alpha \text{ is a } \mu\text{-partition with } H(\alpha) < \infty\} .$$

This theorem is a direct consequence of the following proposition:

3.2.14 Proposition *Let $(X, \mathcal{B}, \mu, \mathcal{T})$ be a m.p.d.s. and let α be a μ-partition with $H(\alpha) < \infty$. Then*

$$h(\alpha, \mathcal{T}) = \sup\{h(\beta, \mathcal{T}) : \beta \preceq \alpha,\ \beta \text{ is a finite } \mu\text{-partition}\} .$$

For the proof of this proposition we need the following lemma:

3.2.15 Lemma *For all μ-partitions α and β with finite entropy*

$$h(\beta, \mathcal{T}) \leq h(\alpha, \mathcal{T}) + \limsup_{n\to\infty} \frac{1}{\lambda_n} H(\beta^{\Lambda_n} \mid \alpha^{\Lambda_n}) \leq h(\alpha, \mathcal{T}) + H(\beta \mid \alpha) .$$

Proof:

$$\begin{aligned} H(\beta^{\Lambda_n} \mid \alpha^{\Lambda_n}) &\leq \sum_{g\in\Lambda_n} H(T^{-g}\beta \mid \alpha^{\Lambda_n}) \quad \text{(Theorem 3.1.8-c)} \\ &\leq \sum_{g\in\Lambda_n} H(T^{-g}\beta \mid T^{-g}\alpha) \quad \text{(Theorem 3.1.8-a)} \\ &= \lambda_n \cdot H(\beta \mid \alpha) \quad \text{(Lemma 3.2.3)} . \end{aligned}$$

Therefore

$$\begin{aligned} H(\beta^{\Lambda_n}) &\leq H(\beta^{\Lambda_n} \vee \alpha^{\Lambda_n}) \quad \text{(Theorem 3.1.8-b)} \\ &= H(\alpha^{\Lambda_n}) + H(\beta^{\Lambda_n} \mid \alpha^{\Lambda_n}) \quad \text{(Corollary 3.1.6-c)} \\ &\leq H(\alpha^{\Lambda_n}) + \lambda_n \cdot H(\beta \mid \alpha) . \end{aligned}$$

After division by λ_n the claim follows as $n \to \infty$. □

3.2.16 Corollary *If α and β are μ-partitions with finite entropy and if $\beta \preceq \alpha$, then $h(\beta, \mathcal{T}) \leq h(\alpha, \mathcal{T})$.*

Proof: $h(\beta, \mathcal{T}) \leq h(\alpha, \mathcal{T}) + H(\beta \mid \alpha) = h(\alpha, \mathcal{T})$ by Lemma 3.1.7. □

Proof of Proposition 3.2.14: Let $\alpha = (A_i : i \in I)$ be a μ-partition with $H(\alpha) = -\sum_{i \in I} \mu(A_i) \log \mu(A_i) < \infty$. Because of the preceding corollary we only have to prove that for given $\epsilon > 0$ there is a finite $\beta \preceq \alpha$ with $h(\beta, \mathcal{T}) \geq h(\alpha, \mathcal{T}) - \epsilon$. So let $\epsilon > 0$. As $H(\alpha) < \infty$, there is a finite subset $J \subseteq I$ such that $|\sum_{i \in I \setminus J} \mu(A_i) \log \mu(A_i)| < \epsilon$. Let $\tilde{A} := X \setminus \bigcup_{i \in J} A_i$. Denote by β the μ-partition of X into the $\mathcal{B}$-measurable sets $\tilde{A}$ and A_i ($i \in J$). Then $\beta \in \mathcal{P}_0$ and in view of Remark 3.1.4

$$
\begin{aligned}
H(\alpha \mid \beta) &= -\sum_{i \in I} \sum_{j \in J} \underbrace{\mu(A_i \cap A_j)}_{=0\ (i \neq j)} \underbrace{\log \mu(A_i \mid A_j)}_{=0\ (i=j)} - \sum_{i \in I} \underbrace{\mu(A_i \cap \tilde{A})}_{=0\ (i \in J)} \log \mu(A_i \mid \tilde{A}) \\
&= -\sum_{i \in I \setminus J} \mu(A_i)(\log \mu(A_i) - \log \mu(\tilde{A})) \\
&< \epsilon .
\end{aligned}
$$

Hence

$$h(\beta, \mathcal{T}) \geq h(\alpha, \mathcal{T}) - H(\alpha \mid \beta) \geq h(\alpha, \mathcal{T}) - \epsilon$$

by Lemma 3.2.15 and the above estimate. As $\epsilon > 0$ was arbitrary, this finishes the proof of the proposition. □

For entropy calculations it is important to know whether the supremum in the definition of $h_{\mathcal{T}}(\mu)$ is attained, and if this is the case, one would like to know for which partitions this is true. To this end we introduce the notion of a generator. (Recall that α^G is the smallest σ-algebra containing all α^{Λ_n}, $n > 0$, when α is a partition of X.)

3.2.17 Definition a) *A μ-partition α of X is a μ-generator for $(X, \mathcal{B}, \mu, \mathcal{T})$, if $\alpha^G = \mathcal{B} \bmod \mu$.*

b) A partition α of X is a generator for the action $\mathcal{T}$ on $(X, \mathcal{B})$, if $\alpha^G = \mathcal{B} \bmod \mu$ for each $\mathcal{T}$-invariant probability measure μ on X.

The following theorem reduces the computation of $h_{\mathcal{T}}(\mu)$ to that of $h(\alpha, \mathcal{T})$ for a fixed generator α.

3.2.18 Theorem (Kolmogorov–Sinai)
If α is a μ-generator for $(X, \mathcal{B}, \mu, \mathcal{T})$ and if $H(\alpha) < \infty$, then $h_{\mathcal{T}}(\mu) = h(\alpha, \mathcal{T})$.

For the proof of this theorem we need the following lemma:

3.2.19 Lemma *If α is a μ-partition and if $\Lambda \subset G$ is finite, then $h(\alpha^\Lambda, \mathcal{T}) = h(\alpha, \mathcal{T})$.*

Proof: Fix $N > 0$ such that $\Lambda \subseteq \Lambda_N$ and observe that $\Lambda_n + \Lambda_N = \Lambda_{n+N-1}$. Then

$$\begin{aligned} h(\alpha^\Lambda, \mathcal{T}) &= \lim_{n\to\infty} \frac{1}{\lambda_n} H((\alpha^\Lambda)^{\Lambda_n}) \\ &\leq \lim_{n\to\infty} \frac{1}{\lambda_n} H(\alpha^{\Lambda_n+\Lambda_N}) = \lim_{n\to\infty} \frac{\lambda_{n+N-1}}{\lambda_n} \frac{1}{\lambda_{n+N-1}} H(\alpha^{\Lambda_{n+N-1}}) \\ &= h(\alpha, \mathcal{T}) \\ &\leq h(\alpha^\Lambda, \mathcal{T}) \quad \text{in view of Corollary 3.2.16.} \end{aligned}$$

□

Proof of Theorem 3.2.18: Let α be a μ-generator of $(X, \mathcal{B}, \mu, \mathcal{T})$ with finite entropy and consider an arbitrary partition $\beta \in \mathcal{P}_0$. For each Λ_n

$$h(\beta, \mathcal{T}) \leq h(\alpha^{\Lambda_n}, \mathcal{T}) + H(\beta \mid \alpha^{\Lambda_n}) = h(\alpha, \mathcal{T}) + H(\beta \mid \alpha^{\Lambda_n})$$

because of Lemmas 3.2.15 and 3.2.19. As

$$\lim_{n\to\infty} H(\beta \mid \alpha^{\Lambda_n}) = H(\beta \mid \mathcal{B}) = 0$$

by Theorem 3.1.10, $h(\beta, \mathcal{T}) \leq h(\alpha, \mathcal{T})$, and as $\beta \in \mathcal{P}_0$ was arbitrary, the claim of the theorem follows. □

3.2.20 Example (Entropy of Bernoulli shifts) (Continuation of Example 3.2.1) Let Σ be a finite set, $\Omega = \Sigma^G$, $G = \mathbb{Z}^d$ or $G = \mathbb{Z}^d_+$, and let $T^g : \Omega \to \Omega$ denote the shift by g. We fix a probability vector $q = (q_\sigma : \sigma \in \Sigma)$ on Σ and denote the corresponding Bernoulli measure on the σ-algebra $\mathcal{B}$ of Ω (generated by the cylinder sets $\mathcal{C}$) by $\mu := q^{\times G}$. For $\sigma \in \Sigma$ let $[\sigma]_0 = \{\omega \in \Omega : \omega_0 = \sigma\}$. Then $\alpha = \{[\sigma]_0 : \sigma \in \Sigma\}$ is a finite generator for the action $\mathcal{T}$ on $(\Omega, \mathcal{B})$ and hence $h_\mathcal{T}(\mu) = h(\alpha, \mathcal{T}) = -\sum_{\sigma\in\Sigma} q_\sigma \log q_\sigma$. If ν is an arbitrary $\mathcal{T}$-invariant measure on $(\Omega, \mathcal{B})$, it follows that $h_\mathcal{T}(\nu) = h(\nu, \alpha, \mathcal{T}) \leq H_\nu(\alpha) \leq \text{card}(\alpha)$ (observe Theorem 3.2.7-b), and the maximal value is attained for the Bernoulli measure with $q_\sigma = |\Sigma|^{-1}$ for all σ. ◇

3.2.21 Exercise Let $(X, \mathcal{B}, \mu, \mathcal{T})$ be a m.p.d.s. For $k \in \mathbb{N}$ let $\mathcal{T}^k := \{T^{k\cdot g} : g \in G\}$. If $G = \mathbb{Z}^d$, we use the same notation for all $k \in \mathbb{Z}$. Prove that

1. $h_{\mathcal{T}^k}(\mu) = k^d \cdot h_\mathcal{T}(\mu)$ $(k \in \mathbb{N})$ and
2. $h_{\mathcal{T}^k}(\mu) = |k|^d \cdot h_\mathcal{T}(\mu)$ $(k \in \mathbb{Z})$, if $G = \mathbb{Z}^d$. ◇

3.2.22 Exercise Let $(X, \mathcal{B}, \mu, \mathcal{T})$ be a m.p.d.s. where $\mathcal{T} = \{T^g : g \in G = \mathbb{Z}^d\}$ with $d \geq 2$. For $g \in G$ denote by $\mathcal{T}_g$ the $\mathbb{Z}$-action $\{T^{k \cdot g} : k \in \mathbb{Z}\}$. Evidently $(X, \mathcal{B}, \mu, \mathcal{T}_g)$ is a m.p.d.s.

Suppose now that there is $g \in G \setminus \{0\}$ such that $h_{\mathcal{T}_g}(\mu) < \infty$. Show that in this case $h_{\mathcal{T}}(\mu) = 0$. *Hint:* Start with the case where g is a multiple of a unit vector $e_j \in G$. Then extend the proof from this special case to general g. ◇

3.2.23 Exercise Let $(X, \mathcal{B}, \mu, T)$ be an ergodic measure preserving $\mathbb{Z}_+$-action. For $A \in \mathcal{B}$ with $\mu(A) > 0$ consider the induced system $(A, \mathcal{B}_A, \mu_A, T_A)$ defined in Section 2.4. Prove *Abramov's formula*

$$h_{T_A}(\mu_A) = \frac{1}{\mu(A)} \cdot h_T(\mu) .$$

Hint: The Shannon–McMillan–Breiman theorem can be of help. ◇

3.2.24 Exercise Let $(X, \mathcal{B}, \mu, \mathcal{T})$ be a m.p.d.s. where $\mathcal{T}$ is a $\mathbb{Z}_+$- or a $\mathbb{Z}$-action. Suppose that $T^{-1}\mathcal{B} = \mathcal{B} \bmod \mu$ and that there exists a finite μ-partition α such that $\alpha_0^\infty = \mathcal{B} \bmod \mu$. Prove that $h_T(\mu) = 0$. Use this to show that that each rotation T_a on S^1 (see Example 2.2.8) has entropy zero. ◇

3.2.25 Exercise Let $X = \Omega = \Sigma^{\mathbb{Z}}$ be a shift space equipped with its Borel σ-algebra $\mathcal{B}$, and denote by T the left shift on Ω. Let $\mathcal{H}_n \subset \mathcal{B}$ be the σ-algebra generated by the coordinate mappings $\omega \mapsto \omega_g$ ($|g| \geq n$) and $\mathcal{H}_\infty := \bigcap_{n \in \mathbb{N}} \mathcal{H}_n$.

a) Prove that for $\mu \in \mathcal{M}(T)$ and any finite μ-partition α of Ω

$$h(\mu, \alpha, T) = 0 \text{ if and only if } \alpha \subseteq \mathcal{H}_\infty \bmod \mu.$$

(In other words: the m.p.d.s. $(X, \mathcal{H}_\infty, \mu, T)$ has entropy zero, and $\mathcal{H}_\infty$ is the largest T-invariant sub-σ-algebra of $\mathcal{B}$ with this property. $\mathcal{H}_\infty$ is called the two-sided *tail-field* of T.) *Hint:* Apply Exercise 3.2.24.

b) Prove that $(X, \mathcal{B}, \mu, T)$ is mixing if $\mathcal{H}_\infty$ is trivial, i.e., if $\mu(H) = 0$ or $\mu(H) = 1$ for each $H \in \mathcal{H}_\infty$. *Warning:* The reverse implication is false, but examples are not easily constructed. ◇

3.2.26 Exercise Let $(X_1, \mathcal{B}_1, \mu_1, \mathcal{T}_1)$ be an extension of $(X_2, \mathcal{B}_2, \mu_2, \mathcal{T}_2)$. Prove that $h_{\mathcal{T}_1}(\mu_1) \geq h_{\mathcal{T}_2}(\mu_2)$ with equality whenever $(X_1, \mathcal{B}_1, \mu_1, \mathcal{T}_1)$ is a natural extension of $(X_2, \mathcal{B}_2, \mu_2, \mathcal{T}_2)$. ◇

3.3 Entropy as a function of the measure

As our motivation for the introduction of the entropies $h(\mu, \alpha, \mathcal{T})$ and $h_{\mathcal{T}}(\mu)$ is the investigation of invariant measures maximizing expressions like $h_{\mathcal{T}}(\mu) + \mu(\psi)$

(see the variational principles in Chapter 1), we are interested not so much in the entropy of a fixed measure preserving dynamical system but instead in entropy of a measurable action $\mathcal{T}$ as a function of the invariant measure μ, i.e., in the *entropy function* $\mu \mapsto h_{\mathcal{T}}(\mu)$.

3.3.1 Definition *For a measurable action $\mathcal{T}$ on a measurable space $(X, \mathcal{B})$ we denote by $\mathcal{M}(\mathcal{T})$ the convex set of all $\mathcal{T}$-invariant probability measures on $\mathcal{B}$.*

3.3.2 Theorem (Affinity of the entropy function)
Let α be a finite partition of X. The functions

$$\mu \mapsto h(\mu, \alpha, \mathcal{T}) \quad \text{and} \quad \mu \mapsto h_{\mathcal{T}}(\mu)$$

are convex affine (i.e., affine under convex combinations), the function $\mu \mapsto H_\mu(\alpha)$ is concave, and for $0 \le a, b \le 1$ with $a + b = 1$ the following holds:

$$H_{a\mu+b\nu}(\alpha^{\Lambda_n}) \le a\, H_\mu(\alpha^{\Lambda_n}) + b\, H_\nu(\alpha^{\Lambda_n}) + \log 2\,. \tag{3.5}$$

Proof: Let $a, b \ge 0$ with $a + b = 1$ and let $\mu, \nu \in \mathcal{M}(\mathcal{T})$. As the function $\varphi(t) = -t \log t$ is concave, for each $B \in \mathcal{B}$

$$\begin{aligned}
0 &\le \varphi(a\mu(B) + b\nu(B)) - a\varphi(\mu(B)) - b\varphi(\nu(B)) \\
&= -a\mu(B) \underbrace{\left[\log(a\mu(B) + b\nu(B)) - \log(a\mu(B))\right]}_{\ge 0} \\
&\quad -b\nu(B) \underbrace{\left[\log(a\mu(B) + b\nu(B)) - \log(b\nu(B))\right]}_{\ge 0} \\
&\quad -\mu(B) a \log a - \nu(B) b \log b \\
&\le -\mu(B) a \log a - \nu(B) b \log b\,,
\end{aligned}$$

and summation over all $B \in \alpha^{\Lambda_n}$ yields

$$0 \le H_{a\mu+b\nu}(\alpha^{\Lambda_n}) - aH_\mu(\alpha^{\Lambda_n}) - bH_\nu(\alpha^{\Lambda_n}) \le -a \log a - b \log b \le \log 2\,.$$

The left hand inequality for $n = 1$ shows already that $\mu \mapsto H_\mu(\alpha)$ is concave, while (3.5) follows from the right hand inequality.

Dividing both inequalities by λ_n we obtain in the limit $n \to \infty$

$$h(a\mu + b\nu, \alpha, \mathcal{T}) = ah(\mu, \alpha, \mathcal{T}) + bh(\nu, \alpha, \mathcal{T})\,,$$

which proves the convex affinity of $\mu \mapsto h(\mu, \alpha, \mathcal{T})$.

Next consider sequences of finite partitions $\alpha_n, \beta_n, \gamma_n$ for which

$$\begin{aligned}
h_{\mathcal{T}}(\mu) &= \lim_{n\to\infty} h(\mu, \alpha_n, \mathcal{T})\,, \\
h_{\mathcal{T}}(\nu) &= \lim_{n\to\infty} h(\nu, \beta_n, \mathcal{T})\,, \\
h_{\mathcal{T}}(a\mu + b\nu) &= \lim_{n\to\infty} h(a\mu + b\nu, \gamma_n, \mathcal{T})\,.
\end{aligned}$$

Let $\delta_n = \alpha_n \vee \beta_n \vee \gamma_n$. Then δ_n is a finite partition which is finer than each of the partitions α_n, β_n, and γ_n so that

$$\begin{aligned}
h_{\mathcal{T}}(\mu) &= \lim_{n\to\infty} h(\mu, \delta_n, \mathcal{T}) \,, \\
h_{\mathcal{T}}(\nu) &= \lim_{n\to\infty} h(\nu, \delta_n, \mathcal{T}) \,, \\
h_{\mathcal{T}}(a\mu + b\nu) &= \lim_{n\to\infty} h(a\mu + b\nu, \delta_n, \mathcal{T}) \,.
\end{aligned}$$

It follows that $\mu \mapsto h_{\mathcal{T}}(\mu)$ is convex affine because the $\mu \mapsto h(\mu, \delta_n, \mathcal{T})$ are. □

Further properties of the entropy function that depend on a topological structure on X are proved in Section 4.1.

Chapter 4

Equilibrium states and pressure

The objects of ergodic theory and entropy theory as outlined in the last chapters are measure preserving dynamical systems, i.e., structures with certain *measurability* properties. In this chapter we make additional topological assumptions on the dynamical systems, essentially that X is a compact metric space and that $\mathcal{T}$ acts continuously on X. There are many textbooks treating various aspects of this theory in the case $d = 1$ (e.g., [6], [14], [60]). A good reference for more general group actions (including $G = \mathbb{Z}^d$) is [42]. Interesting generalizations can be found in [45].

4.1 Pressure

For systems on finite lattices the notion of *pressure* was already introduced in Section 1.2. In this section we extend that notion to a broad class of dynamical systems including, in particular, the shift systems from Example 3.2.1. We start with the definition of the basic objects of our study in this chapter:

4.1.1 Definition *The pair $(X, \mathcal{T})$ is called a topological dynamical system (t.d.s.), if*

1. *X is a compact metrizable space and*
2. *$\mathcal{T}$ is a continuous action of $G = \mathbb{Z}^d$ or $G = \mathbb{Z}^d_+$ on X (i.e., $T^g : X \to X$ is continuous for all $g \in G$).*

In this setting we denote by $\mathcal{B}$ the Borel σ-algebra of X and note that $\mathcal{T}$ acts measurably on the measurable space $(X, \mathcal{B})$.

Recall that $\mathcal{M}(\mathcal{T})$ is the set of all $\mathcal{T}$-invariant Borel probability measures on X and that $\mathcal{M}(\mathcal{T})$ is a convex subspace of the set $\mathcal{M}$ of all Borel probability measures on X. From a statistical mechanics point of view the measures in $\mathcal{M}$ or $\mathcal{M}(\mathcal{T})$ might be called states again as in Section 1.1. The following two lemmas together generalize the Krylov–Bogolubov theorem.

4.1.2 Lemma *Let $m_n \in \mathcal{M}$, $\mu_n := \frac{1}{\lambda_n}\sum_{g\in\Lambda_n} T^g m_n$, and suppose that $\mu = \lim_{i\to\infty}\mu_{n_i}$. Then $\mu \in \mathcal{M}(\mathcal{T})$.*

Proof: Fix $f \in C(X)$ and $g \in G$. Then

$$\begin{aligned}
|\mu(f - f\circ T^g)| &= \lim_{i\to\infty} |\mu_{n_i}(f - f\circ T^g)| \\
&\leq \lim_{i\to\infty} \frac{1}{\lambda_{n_i}} \sum_{h\in\Lambda_{n_i}\Delta(\Lambda_{n_i}+g)} |m_{n_i}(f\circ T^h)| \\
&\leq \lim_{i\to\infty} 2\,\frac{\lambda_{n_i} - \lambda_{n_i-|g|}}{\lambda_{n_i}}\,\|f\|_\infty \\
&= 0
\end{aligned}$$

where $|g| := \max_i |g_i| \leq n$. As $g \in G$ and $f \in C(X)$ were arbitrary, we conclude that $\mu \in \mathcal{M}(\mathcal{T})$. □

4.1.3 Lemma *$\mathcal{M}(\mathcal{T})$ is a nonempty, compact, convex subset of $\mathcal{M}$.*

Proof: As the set $\mathcal{M}$ of all Borel probability measures on X is compact in the weak topology (see A.4.30), $\mathcal{M}(\mathcal{T}) \neq \emptyset$ in view of the preceding lemma. So we have to show that $\mathcal{M}(\mathcal{T})$ is closed in $\mathcal{M}$. To this end let $\mu_n \in \mathcal{M}(\mathcal{T})$, $\mu \in \mathcal{M}$, and suppose that $\mu_n \to \mu$ weakly. Then for each $f \in C(X)$ and $T \in \mathcal{T}$ the function $f\circ T$ is also in $C(X)$ and

$$\mu(f\circ T) = \lim_{n\to\infty}\mu_n(f\circ T) = \lim_{n\to\infty}\mu_n(f) = \mu(f)\,.$$

It follows that $\mu \in \mathcal{M}(\mathcal{T})$. □

Motivated by the considerations in Chapter 1 we are going to define the *pressure* of a *local energy function* $\psi : X \to \mathbb{R}$ by

$$p(\psi) := \sup_{\mu\in\mathcal{M}(\mathcal{T})} (h(\mu) + \mu(\psi))$$

(cf. Equation (1.4)), ask for existence and uniqueness of measures $\mu \in \mathcal{M}(\mathcal{T})$ for which this supremum is attained, and attempt to construct such measures explicitly. It turns out that for a precise mathematical formulation of this programme the local energy functions should be upper semicontinuous. As, even more importantly, upper semicontinuity will also turn out to be a key property of the entropy function, we collect here some facts about upper semicontinuous functions.

4.1.4 Definition *Let Y be a topological space. A function $f : Y \to [-\infty, \infty)$ is upper semicontinuous, if $\{y \in Y : f(y) < c\}$ is open for each $c \in \mathbb{R}$. The class of all upper semicontinuous $f : Y \to [-\infty, \infty)$ is denoted by $USC(Y)$.*

4.1.5 Lemma *Let Y be a compact, metrizable topological space. Then $\sup f < \infty$ for each $f \in USC(Y)$, and the following assertions are equivalent:*

1. *$f \in USC(Y)$.*
2. *$\{y \in Y : f(y) \geq c\}$ is closed for each $c \in \mathbb{R}$.*
3. *If $y, y_n \in Y$ and $\lim_{n\to\infty} y_n = y$, then $\limsup_{n\to\infty} f(y_n) \leq f(y)$.*
4. *There is a decreasing sequence of continuous functions $f_n : Y \to \mathbb{R}$ such that $f(y) = \lim_{n\to\infty} f_n(y)$ for each $y \in Y$.*

Proof: The assertion $\sup f < \infty$, the equivalence of 1, 2 and 3 and the implication from 4 to 3 are obvious. We prove that 2 implies 4. For $t \in \mathbb{R}$ we use the shorthand notation $\{f \geq t\}$ for $\{y \in Y : f(y) \geq t\}$. Observe that $\{f \geq t\}$ is closed for each $t \in \mathbb{R}$ by assumption. Let ρ be a metric for the topology on Y. For $n \in \mathbb{N}$, $t \in \mathbb{R}$, and $y \in Y$ let

$$\kappa_n(t, y) := \max\{0, 1 - n \cdot \rho(y, \{f \geq t\})\}$$

where $\rho(y, A) := \min\{\rho(y, a) : a \in A\}$ for each closed subset $A \subseteq Y$ and $\rho(y, \emptyset) := \infty$ by convention. Then $0 \leq \kappa_n \leq 1$, and for each n and t the function $y \mapsto \kappa_n(t, y)$ is Lipschitz-continuous with Lipschitz constant n. Observe that for $y \in Y$ the following hold:

- ▷ $f(y) \geq t \implies \rho(y, \{f \geq t\}) = 0 \implies \kappa_n(t, y) = 1$ and
- ▷ $f(y) < t \implies \rho(y, \{f \geq t\}) > 0 \implies \kappa_n(t, y) \searrow 0$ as $n \to \infty$.

As f is bounded from above, there is $M > 0$ such that $f \leq M$, and if $f(y) > -n$, then

$$f(y) = \int_{-n}^{M} 1_{\{f(y) \geq t\}}\, dt - n\,.$$

Let

$$f_n(y) := \int_{-n}^{M} \kappa_n(t, y)\, dt - n\,.$$

Then all f_n are continuous (in fact, Lipschitz-continuous with Lipschitz constant $n(n + M)$). The sequence $(f_n)_{n>0}$ decreases, since

$$\begin{aligned} & f_n(y) - f_{n+1}(y) \\ = \; & \underbrace{\int_{-n}^{M} (\kappa_n(t, y) - \kappa_{n+1}(t, y))\, dt}_{\geq 0} - \underbrace{\int_{-(n+1)}^{-n} \kappa_{n+1}(t, y)\, dt}_{\leq 1} - n + (n + 1) \\ \geq \; & 0 \end{aligned}$$

for all y.

If $f(y) > -\infty$, then for all $n > -f(y)$

$$\begin{aligned} f_n(y) &= \int_{-n}^{f(y)} \kappa_n(t,y)\,dt - n + \int_{f(y)}^{M} \kappa_n(t,y)\,dt \\ &= (f(y)+n) - n + \int_{f(y)}^{M} \kappa_n(t,y)\,dt \searrow f(y)\,. \end{aligned}$$

If $f(y) = -\infty$, then $\kappa_n(t,y) \searrow 0$ for all t. Hence, if $n > K$, then

$$f_n(y) = \underbrace{\int_{-n}^{-K} (\kappa_n(t,y) - 1)\,dt}_{\leq 0} + \underbrace{\int_{-K}^{M} \kappa_n(t,y)\,dt}_{\searrow 0\ (n\to\infty)} - K\,.$$

As this holds for all $K > 0$, we have $f_n(y) \searrow f(y)$ in this case, too. □

4.1.6 Remarks
a) If $F \subseteq USC(Y)$ and if $g(y) = \inf_{f\in F} f(y)$ for all $y \in Y$, then $g \in USC(Y)$, because $\{y : g(y) \geq c\} = \bigcap_{f\in F}\{y : f(y) \geq c\}$ for each $c \in \mathbb{R}$.

b) $USC(Y)$ is closed under addition and multiplication by positive scalars. So it is a convex set, although it is difficult to consider it as subset of a real vector space, because we allow upper semicontinuous functions to take the value $-\infty$.

◇

4.1.7 Corollary *Let $\mu_n, \mu \in \mathcal{M}$ and $\psi \in USC(X)$. If $\mu = \lim_{n\to\infty} \mu_n$, then*

$$\limsup_{n\to\infty} \mu_n(\psi) \leq \mu(\psi)\,.$$

Proof: Because of Lemma 4.1.5 there is a decreasing sequence of functions $\psi_k \in C(X)$ such that $\lim_{k\to\infty} \psi_k(x) = \psi(x)$ for all $x \in X$. Therefore $\limsup_{n\to\infty} \mu_n(\psi) \leq \lim_{n\to\infty} \mu_n(\psi_k) = \mu(\psi_k)$ for all k. As $\mu(\psi) = \inf_k \mu(\psi_k)$ by the monotone convergence theorem (A.4.19), the corollary follows. □

From now on we assume that (X, T) is a topological dynamical system, in particular that X is a compact metrizable space. Therefore each $f \in USC(X)$ is bounded from above.

4.1.8 Definition *1. Let $\psi \in USC(X)$, $\inf \psi > -\infty$. The pressure of ψ is*

$$p(\psi) := \sup_{\mu\in\mathcal{M}(T)} (h(\mu) + \mu(\psi))\,. \tag{4.1}$$

In particular $p(\psi)$ is defined for each $\psi \in C(X)$.

2. The number $p(0) = \sup_{\mu\in\mathcal{M}(T)} h(\mu)$ is called the topological entropy of T.

3. If $p(0) < \infty$, then the definition of pressure extends to all $\psi \in USC(X)$.

4.1.9 Remarks
a) $p(\psi) = -\infty$ if and only if $\mu(\psi) = -\infty$ for all $\mu \in \mathcal{M}(\mathcal{T})$.

b) If $p(0) = \infty$, then $p(\psi) = \infty$ for all $\psi \in USC(X)$ that are bounded from below, because

$$p(\psi) = \sup_{\mu \in \mathcal{M}(\mathcal{T})} (h(\mu) + \mu(\psi)) \geq \sup_{\mu \in \mathcal{M}(\mathcal{T})} h(\mu) + \inf \psi = p(0) + \inf \psi = \infty .$$

c) Similarly, if $p(0) < \infty$, then $p(\psi) < \infty$ for all $\psi \in USC(X)$. ◇

4.1.10 Theorem (Basic properties of the pressure function)
Suppose that $p(0) < \infty$. Then

a) $p : USC(X) \to [-\infty, \infty)$ is convex and isotonic (i.e., $p(\phi) \leq p(\psi)$ if $\phi \leq \psi$).

b) For $\psi, \phi \in USC(X)$

$$p(\psi) + \inf \phi \leq p(\psi + \phi) \leq p(\psi) + \sup \phi .$$

In particular, $|p(\psi + \phi) - p(\psi)| \leq \|\phi\|_\infty$ if the difference is well defined.

c) For $\psi \in USC(X)$, $\phi \in C(X)$ and $g \in G$

$$p(\psi + \phi \circ T^g - \phi) = p(\psi) \quad \text{(and hence } p(\psi + \phi \circ T^g) = p(\psi + \phi)\text{)} .$$

Proof: By definition, p is the supremum of the affine, isotonic functionals $p_\mu : \psi \mapsto h(\mu) + \mu(\psi)$ ($\mu \in \mathcal{M}(\mathcal{T})$). Therefore p is isotonic and convex. As $p_\mu(\psi) + \inf \phi \leq p_\mu(\psi + \phi) \leq p_\mu(\psi) + \sup \phi$ for all $\mu \in \mathcal{M}(\mathcal{T})$, the same inequalities hold for p. Finally

$$p_\mu(\psi + \phi \circ T^g - \phi) = h(\mu) + \mu(\psi) + \mu(\phi \circ T^g) - \mu(\phi) = h(\mu) + \mu(\psi) = p_\mu(\psi)$$

for $\psi \in USC(X)$ and $\phi \in C(X)$, and this identity carries over to $p = \sup_\mu p_\mu$. □

As an immediate corollary we have the following analogue to Lemma 1.1.2:

4.1.11 Corollary *For $\psi \in C(X)$ and $\beta \in \mathbb{R}$ let $Z(\beta) := e^{p(-\beta\psi)}$ be the partition function of ψ. The function $\beta \mapsto \log Z(\beta) = p(-\beta\psi)$ is convex and Lipschitz-continuous with Lipschitz constant $\|\psi\|_\infty$.*

4.1.12 Remark In Chapter 6 an extension of Theorem 4.1.10-c to more general functions ϕ is needed. So assume that $\phi : X \to [-\infty, \infty]$ is measurable and that $(\phi \circ T^g - \phi)$ is bounded from above or from below outside the set $A := \{x : \phi(x) = \phi(T^g x) = \pm\infty\}$ for some $g \in G$. (On A the difference $(\phi \circ T^g - \phi)$ is not well defined.) The subsequent lemma shows that in this situation $(\phi \circ T^g - \phi) \in L^1_\mu$ and $\mu(\phi \circ T^g - \phi) = 0$ for all $\mu \in \mathcal{M}(\mathcal{T})$ such that $\mu(A) = 0$. In particular, if $A = \emptyset$, then $p(\psi + \phi \circ T^g - \phi) = p(\psi)$ is well defined. ◇

4.1.13 Lemma *Let $(X, \mathcal{F}, \mu)$ be a probability space, $T : X \to X$ a measurable transformation leaving the measure μ invariant, and $\phi : X \to [-\infty, \infty]$ a measurable function such that $\mu\{x : \phi(x) = \phi(Tx) = \pm\infty\} = 0$. If the function $(\phi \circ T - \phi)$ is bounded from below by some $u \in L^1_\mu$, then $(\phi \circ T - \phi) \in L^1_\mu$ and*

$$\int (\phi \circ T - \phi)\, d\mu = 0\,.$$

Proof: Let $\phi_n := \max(\min(\phi, n), -n)$. Then

$$\begin{aligned} 0 \le \phi_n \circ T - \phi_n \le \phi \circ T - \phi \quad &\text{on the set} \quad \{\phi \circ T - \phi \ge 0\} \quad \text{and} \\ 0 \ge \phi_n \circ T - \phi_n \ge \phi \circ T - \phi \quad &\text{on the set} \quad \{\phi \circ T - \phi \le 0\}\,. \end{aligned}$$

Therefore $(\phi_n \circ T - \phi_n)_{n>0}$ is a sequence of bounded functions with common integrable minorant $\min(u, 0)$ and converging to $(\phi \circ T - \phi)$ μ-a.e. Observing the T-invariance of μ it thus follows from Fatou's lemma (A.4.18) that

$$\int (\phi \circ T - \phi)\, d\mu \le \liminf_{n\to\infty} \int (\phi_n \circ T - \phi_n)\, d\mu = 0\,.$$

Hence $(\phi \circ T - \phi) \in L^1_\mu$. Because of $|\phi_n \circ T - \phi_n| \le |\phi \circ T - \phi|$, the dominated convergence theorem (A.4.20) finally yields $\int (\phi \circ T - \phi)\, d\mu = 0$. □

4.2 Equilibrium states and the entropy function

For $\psi \in USC(X)$ we defined the pressure $p(\psi) := \sup_{\mu\in\mathcal{M}(T)}(h(\mu) + \mu(\psi))$. (Henceforth we often call any $\psi \in USC(X)$ a *local energy function.*) In this section we define equilibrium states as those measures for which this supremum is attained, and we show that such measures always exist, if the entropy function is upper semicontinuous.

4.2.1 Definition *A measure $\mu \in \mathcal{M}(T)$ is an equilibrium state for $\psi \in USC(X)$ if*

$$p(\psi) = h(\mu) + \mu(\psi)\,.$$

The set of all equilibrium states for ψ is denoted by $ES(\psi, T)$. In the case $\psi = 0$ an equilibrium state μ satisfies

$$h(\mu) = p(0) = \sup_{\nu\in\mathcal{M}(T)} h(\nu)$$

and is called a measure of maximal entropy for T.

4.2.2 Example (Bernoulli measures as equilibrium states) Let $(\Sigma^G, \mathcal{T})$ be the shift over a finite alphabet Σ with generator $\alpha = \{[\sigma]_0 : \sigma \in \Sigma\}$, see Example 3.2.20. An appropriate metric that makes $\Omega = \Sigma^G$ a compact space is $d(\omega, \omega') := 2^{-n(\omega,\omega')}$ where $n(\omega, \omega') := \sup\{n \in \mathbb{N} : \omega_g = \omega'_g \ \forall g \in \Lambda_n\}$. In fact, cylinder sets $[\sigma_g]_{g\in\Lambda_n} \in \mathcal{C}_n$ (see Example 2.2.10 for the definition) are open and closed in the topology determined by d. More exactly, a set $C \in \mathcal{C}_n$ is at the same time the closed ball of radius 2^{-n} and the open ball of radius $2^{-(n-1)}$ around each of its elements.

As each $\mathcal{C}_n$ is finite with $\bigcup_{C\in\mathcal{C}_n} C = \Omega$ and as $\operatorname{diam}(C) = 2^{-n}$ for all $C \in \mathcal{C}_n$, the metric space (Ω, d) is totally bounded. Furthermore, a straightforward application of the pigeon-hole principle shows that each Cauchy sequence in Ω converges. Hence (Ω, d) is compact.

Let ψ be a local energy function that depends only on the zero coordinate, $\psi(\omega) = u(\omega_0)$ for some function $u : \Sigma \to \mathbb{R}$. Then ψ is certainly continuous. Fix some probability vector $q = (q_\sigma : \sigma \in \Sigma)$ on Σ and denote the corresponding Bernoulli measure by μ. Let $\nu \in \mathcal{M}(\mathcal{T})$ be arbitrary with $\nu([\sigma]_0) = q_\sigma$ for all $\sigma \in \Sigma$. Then $\nu(\psi) = \mu(\psi)$ and

$$h(\nu) = \lim_{n\to\infty} \frac{1}{\lambda_n} H_\nu(\alpha^{\Lambda_n}) \leq H_\nu(\alpha) = H(q) = h(\mu)$$

by Corollary 3.1.6, Lemma 3.2.3, and Example 3.2.20 where we use the notation $H(q) = -\sum_{\sigma\in\Sigma} q_\sigma \log q_\sigma$ as in Section 1.1. Therefore

$$h(\nu) + \nu(\psi) \leq h(\mu) + \mu(\psi) = H(q) + \sum_{\sigma\in\Sigma} q_\sigma u(\sigma)$$

and it follows that

$$p(\psi) = \sup\left\{ H(q) + \sum_{\sigma\in\Sigma} q_\sigma u(\sigma) : q \text{ a probability vector on } \Sigma \right\}.$$

In view of Remark 1.1.9, $p(\psi) = \sum_{\sigma\in\Sigma} e^{u(\sigma)}$ and the supremum is attained for the probability vector $\tilde{q}$ with $\tilde{q}_\sigma = e^{u(\sigma)-p(\psi)}$. Therefore the corresponding Bernoulli measure $\tilde{\mu}$ is an equilibrium state for ψ. ◇

While the uniqueness problem for equilibrium states turns out to be rather delicate and (partial) answers to it are postponed to Chapter 5, their existence can be proved under quite general assumptions. A first step is the following theorem:

4.2.3 Theorem (Existence of equilibrium states)
If the entropy function $h : \mathcal{M}(\mathcal{T}) \to [0, \infty]$ is upper semicontinuous, then $p(0) < \infty$ and for each $\psi \in USC(X)$ the set $ES(\psi, \mathcal{T})$ is a nonempty, compact, convex subset of $\mathcal{M}(\mathcal{T})$.

Proof: Observe first that $p(0) = \sup_{\mu\in\mathcal{M}(\mathcal{T})} h(\mu) < \infty$ because the upper semicontinuous function h is bounded by Lemma 4.1.5. As the entropy function is affine by Theorem 3.3.2, the sets $F_k := \{\mu \in \mathcal{M}(\mathcal{T}) : h(\mu) + \mu(\psi) \geq p(\psi) - \frac{1}{k}\}$ are convex, compact, and nonempty. Hence the same is true for the set

$$\begin{aligned}\bigcap_{k>0} F_k &= \{\mu : h(\mu) + \mu(\psi) \geq p(\psi)\} = \{\mu : h(\mu) + \mu(\psi) = p(\psi)\} \\ &= ES(\psi, \mathcal{T}).\end{aligned}$$

□

The next theorem provides a sufficient condition for the upper semicontinuity of the entropy function:

4.2.4 Theorem (Upper semicontinuity of the entropy function)

a) *Let $\mu_n, \mu \in \mathcal{M}(\mathcal{T})$ and suppose that $\lim_{n\to\infty} \mu_n = \mu$. Let α be a finite partition satisfying*

$$\lim_{m\to\infty} \frac{1}{\lambda_m} \left(\sum_{A\in\alpha^{\Lambda_m}} \mu(\bar{A} \setminus A)\,|\log \mu(A)| + \log a_m \right) = 0 \tag{4.2}$$

where $\bar{A}$ is the topological closure of A and

$$a_m := |\{A \in \alpha^{\Lambda_m} : \mu(\bar{A}) > e^{-1}\}| + 1\,.$$

Then

$$\limsup_{n\to\infty} h(\mu_n, \alpha, \mathcal{T}) \leq \limsup_{m\to\infty} \frac{1}{\lambda_m} \left(\limsup_{n\to\infty} H_{\mu_n}(\alpha^{\Lambda_m}) \right) \leq h(\mu, \alpha, \mathcal{T})\,. \tag{4.3}$$

b) *Suppose that for each sequence $\mu_n \in \mathcal{M}(\mathcal{T})$ converging to some $\mu \in \mathcal{M}(\mathcal{T})$ there exists a finite partition α which is a μ_n-generator for all n (see Definition 3.2.17) and satisfies (4.2). Then the entropy function h is upper semicontinuous.*

c) *Assumption (4.2) is in particular satisfied if $\mu(\partial A) = 0$ for all $A \in \alpha$.*

Proof:

a) Let μ_n, μ and α be as in the assumptions of the theorem. As $1_{\bar{A}}$ is upper semicontinuous, $\limsup_{n\to\infty} \mu_n(A) \leq \mu(\bar{A})$ for all $A \in \alpha^{\Lambda_m}$ by Corollary 4.1.7. Let $\mathcal{Z}_m := \{A \in \alpha^{\Lambda_m} : \mu(\bar{A}) \leq e^{-1}\}$ and $\mathcal{Z}'_m := \alpha^{\Lambda_m} \setminus \mathcal{Z}_m$. As $\varphi(t) = -t \log t$ is increasing for $0 \leq t \leq e^{-1}$, it follows that $\limsup_{n\to\infty} \varphi(\mu_n(A)) \leq \varphi(\mu(\bar{A}))$ for all $A \in \mathcal{Z}_m$. Hence

$$\limsup_{n\to\infty} H_{\mu_n}(\alpha^{\Lambda_m}) \leq \sum_{A\in\mathcal{Z}_m} \varphi(\mu(\bar{A})) + \limsup_{n\to\infty} \sum_{A\in\mathcal{Z}'_m} \varphi(\mu_n(A))$$

$$\leq -\sum_{A\in\mathcal{Z}_m} \mu(\bar{A})\log\mu(A) + \log\left(|\mathcal{Z}'_m|+1\right)$$

$$\leq H_\mu(\alpha^{\Lambda_m}) - \sum_{A\in\alpha^{\Lambda_m}} \mu(\bar{A}\setminus A)\log\mu(A) + \log\left(|\mathcal{Z}'_m|+1\right) .$$

In view of (4.2) this yields

$$\limsup_{m\to\infty}\frac{1}{\lambda_m}\left(\limsup_{n\to\infty} H_{\mu_n}(\alpha^{\Lambda_m})\right) \leq h(\mu,\alpha,\mathcal{T}) ,$$

and as $h(\mu_n,\alpha,\mathcal{T}) = \inf_{m>0}\frac{1}{\lambda_m}H_{\mu_n}(\alpha^{\Lambda_m})$, (4.3) follows at once.

b) Given a sequence of measures $\mu_n \in \mathcal{M}(\mathcal{T})$ converging weakly to $\mu \in \mathcal{M}(\mathcal{T})$ we have to show that $\limsup_{n\to\infty} h(\mu_n) \leq h(\mu)$. Choose α as in the assumptions of the theorem. It follows from (4.3) that

$$\limsup_{n\to\infty} h(\mu_n) = \limsup_{n\to\infty} h(\mu_n,\alpha,\mathcal{T}) \leq h(\mu,\alpha,\mathcal{T}) \leq h(\mu) .$$

c) As for each finite subset $\Lambda \subseteq G$ and each $B \in \alpha^\Lambda$

$$\partial B \subseteq \bigcup_{g\in\Lambda}\bigcup_{A\in\alpha} \partial(T^{-g}A) \subseteq \bigcup_{g\in\Lambda}\bigcup_{A\in\alpha} T^{-g}(\partial A) ,$$

the measure of the boundaries ∂B is

$$\mu(\partial B) \leq \sum_{g\in\Lambda}\sum_{A\in\alpha} \mu(\partial A) = 0 .$$

Hence $\mu(\bar{A}\setminus A) = 0$ for each $A \in \alpha^{\Lambda_m}$ and $a_m \leq 2$. □

4.2.5 Corollary *Let $\alpha = \{A_1,\dots,A_r\}$ be a finite generator for the action $\mathcal{T}$ on $(X,\mathcal{B})$ and assume that $\mu(\partial A) = 0$ for all $\mu \in \mathcal{M}(\mathcal{T})$ and all $A \in \alpha$. Then the entropy function h is upper semicontinuous (and for each $\psi \in USC(X)$ there is at least one equilibrium state by Theorem 4.2.3).*

4.2.6 Example (Shift systems) The last corollary applies in particular to shift systems $(\Sigma^G,\mathcal{T})$ because the sets $[\sigma]_0$ ($\sigma \in \Sigma$) are open and closed, hence $\partial[\sigma]_0 = \emptyset$ for each $\sigma \in \Sigma$, see Example 4.2.2. Example 3.2.20 shows more directly that $p(0) = \log|\Sigma|$ and $ES(0) \neq \emptyset$. ◇

4.2.7 Remark (Closed subshifts) Let $(\Sigma^G,\mathcal{T})$ be a shift system as in the previous example. If $X \subseteq \Sigma^G$ is closed and $\mathcal{T}$-invariant, then $(X,\mathcal{T}_{|X})$ is also a t.d.s. and there is a natural one-to-one correspondence between $\mathcal{M}(\mathcal{T}_{|X})$ and $\{\mu \in \mathcal{M}(\mathcal{T}) : \mu(X) = 1\}$. For $\psi \in C(X)$ define $\psi^* : \Sigma^G \to [-\infty,\infty)$ by $\psi^*(\omega) = \psi(\omega)$ if $\omega \in X$ and $\psi^*(\omega) = -\infty$ otherwise. Then $\psi^* \in USC(X)$ because $X \subset \Sigma^G$ is closed (see Lemma 4.1.5), and for $\mu \in \mathcal{M}(\mathcal{T})$ the following holds: $\mu(\psi^*) > -\infty$ if and only if $\mu(X) = 1$. Therefore

$$\mu \in ES(\psi^*, \mathcal{T}) \quad \text{if and only if} \quad \mu(X) = 1 \text{ and } \mu \in ES(\psi, \mathcal{T}_{|X}) .$$

◇

4.2.8 Remark It is not difficult to construct a t.d.s. of positive topological entropy for which the entropy function is not upper semicontinuous at a measure of entropy zero, see [60, p. 184]. This system does possess a measure of maximal entropy, however. A more elaborate construction is needed to produce a continuous map of a compact metric space with finite topological entropy which has no measure of maximal entropy, see [27]. ◇

The next theorem shows that the upper semicontinuity of the entropy function is not merely a convenient technical assumption but that it is equivalent to a variational equation dual to (4.1).

4.2.9 Theorem (Dual variational principle)
Suppose that $p(0) < \infty$. Then the entropy function h is upper semicontinuous if and only if

$$h(\mu) = \inf_{\psi \in C(X)} (p(\psi) - \mu(\psi)) \quad \text{for all } \mu \in \mathcal{M}(\mathcal{T}). \tag{4.4}$$

Proof: As the maps $p_\psi : \mathcal{M}(\mathcal{T}) \to \mathbb{R}$, $\mu \mapsto (p(\psi) - \mu(\psi))$ are continuous for each $\psi \in C(X)$, it follows from Remark 4.1.6-a that the validity of (4.4) is sufficient for the upper semicontinuity of the entropy function. Furthermore, "$\leq$" in (4.4) always holds by definition of $p(\psi)$. It remains to prove the "$\geq$"-direction in (4.4) assuming that h is upper semicontinuous.

To this end fix $\mu \in \mathcal{M}(\mathcal{T})$ and $h > h(\mu)$ and consider the set

$$C := \{(\nu, t) \in \mathcal{M}(\mathcal{T}) \times \mathbb{R} : 0 \leq t \leq h(\nu)\} .$$

As the entropy function h is convex affine (see Theorem 3.3.2), the set C is convex, and because of the upper semicontinuity of h it is a closed set with respect to the product topology on $\mathcal{M}(\mathcal{T}) \times [0, p(0)]$ (see Lemma 4.1.5). As $p(0) < \infty$ the latter set is compact so that C is a compact convex set, and $(\mu, h) \notin C$ because $h > h(\mu)$. Hence there are $\psi_1, \ldots, \psi_n \in C(X)$ and $\epsilon > 0$ such that the open neighbourhood

$$U := \{(\nu, t) \in \mathcal{M}(\mathcal{T}) \times \mathbb{R} : |t - h| < \epsilon;\ |\nu(\psi_i) - \mu(\psi_i)| < \epsilon\ \forall i\}$$

of (μ, h) is disjoint from C. In order to separate (μ, h) and C by some hyperplane consider the continuous and affine functional

$$F : \mathcal{M}(\mathcal{T}) \times \mathbb{R} \to \mathbb{R}^{n+1},\ (\nu, t) \mapsto (t, \nu(\psi_1), \ldots, \nu(\psi_n))$$

that maps C onto the convex compact set $F(C) \subseteq \mathbb{R}^{n+1}$. As $U \cap C = \emptyset$ and $U = F^{-1}(B_\epsilon(F(\mu, h)))$, we have also $F(U) \cap F(C) = \emptyset$ so that $F(\mu, h) \notin F(C)$. Therefore $F(\mu, h)$ and $F(C)$ can be strictly separated by some hyperplane in $\mathbb{R}^{n+1}$ (see Appendix, A.2), i.e., there are $a, a_1, \ldots, a_n \in \mathbb{R}$ such that

$$at + \sum_{i=1}^{n} a_i \nu(\psi_i) < ah + \sum_{i=1}^{n} a_i \mu(\psi_i) \quad \text{for all } (\nu, t) \in C.$$

Since this inequality holds in particular for $(\nu, t) = (\mu, h(\mu)) \in C$, it follows that $ah(\mu) < ah$ and hence $a > 0$.

Let $\bar{\psi} = \frac{1}{a} \sum_i a_i \psi_i$. Then $\bar{\psi} \in C(X)$ and

$$p(\bar{\psi}) = \sup_{\nu \in \mathcal{M}(\mathcal{T})} \left(h(\nu) + \nu(\bar{\psi}) \right) = \sup_{(\nu,t) \in C} \left(t + \nu(\bar{\psi}) \right) \leq h + \mu(\bar{\psi}),$$

and as $h > h(\mu)$ was arbitrary, we conclude that

$$h(\mu) \geq p(\bar{\psi}) - \mu(\bar{\psi}) \geq \inf_{\psi \in C(X)} (p(\psi) - \mu(\psi)) .$$

□

4.2.10 Remark For each probability measure μ on X, and not only for $\mathcal{T}$-invariant measures, the right hand side of (4.4) is a well defined number in $[-\infty, p(0)]$. Therefore it may be considered as an extension of the notion of entropy to non-invariant probability measures. However, as such it is not very useful, because $\inf_{\psi \in C(X)} (p(\psi) - \mu(\psi)) = -\infty$ if μ is not $\mathcal{T}$-invariant. In fact, if μ is not $\mathcal{T}$-invariant, then there are $T \in \mathcal{T}$ and $\phi \in C(X)$ such that $\mu(\phi \circ T - \phi) > 0$. In view of Theorem 4.1.10 it follows that for each $t > 0$

$$\inf_{\psi \in C(X)} (p(\psi) - \mu(\psi)) \leq p(t\phi \circ T - t\phi) - \mu(t\phi \circ T - t\phi) = p(0) - t\mu(\phi \circ T - \phi)$$

which proves the claim. ◇

The next theorem describes a kind of continuous dependence of the set $ES(\psi, \mathcal{T})$ on the function ψ.

4.2.11 Theorem (Continuous dependence of equilibrium states)
Suppose that the entropy function h is upper semicontinuous. Consider $\psi, \phi_n \in USC(X)$ and $t_n \in (-1, 1)$ such that $\lim_{n\to\infty} t_n = 0$ and $\lim_{n\to\infty} \|\phi_n\|_\infty = 0$. Let $\mu_n \in ES((1 + t_n)\psi + \phi_n, \mathcal{T})$, $n > 0$.

a) If $(\mu_n)_{n>0}$ converges weakly to some $\mu \in \mathcal{M}(\mathcal{T})$, then $\mu \in ES(\psi, \mathcal{T})$.

b) If ψ has a unique equilibrium state μ, then $\lim_{n\to\infty} \mu_n = \mu$.

Proof:

a) If $p(\psi) = -\infty$, then $ES(\psi, \mathcal{T}) = \mathcal{M}(\mathcal{T})$ so that $\mu \in ES(\psi, \mathcal{T})$. Otherwise $p(\psi) > -\infty$, a fact that we use later. Observe that

$$\begin{aligned} & p((1+t_n)\psi + \phi_n) \\ = & \sup_{\mu \in \mathcal{M}(\mathcal{T})} \Big(h(\mu) + \mu((1+t_n)\psi + \phi_n) \Big) \\ = & \sup_{\mu \in \mathcal{M}(\mathcal{T})} \Big((1+t_n)(h(\mu) + \mu(\psi)) - t_n h(\mu) + \mu(\phi_n) \Big) \\ \geq & (1+t_n)p(\psi) - |t_n| p(0) - \|\phi_n\|_\infty . \end{aligned} \tag{4.5}$$

Recall that $p(0) < \infty$ by Theorem 4.2.3. As h is upper semicontinuous this implies

$$\begin{aligned} & h(\mu) + \mu(\psi) \\ \geq & \limsup_{n\to\infty} h(\mu_n) + \limsup_{n\to\infty} \mu_n(\psi) \\ \geq & \limsup_{n\to\infty} \Big(h(\mu_n) + \mu_n((1+t_n)\psi + \phi_n) - \mu_n(t_n\psi + \phi_n) \Big) \\ \geq & \limsup_{n\to\infty} \Big(p((1+t_n)\psi + \phi_n) - |t_n \mu_n(\psi)| - \|\phi_n\|_\infty \Big) \\ \geq & \limsup_{n\to\infty} \Big((1+t_n)p(\psi) - |t_n| p(0) - 2\|\phi_n\|_\infty - |t_n \mu_n(\psi)| \Big) \quad \text{by (4.5)} \\ \geq & p(\psi) - \limsup_{n\to\infty} |t_n \mu_n(\psi)| . \end{aligned}$$

It remains to show that $\limsup_{n\to\infty} |\mu_n(\psi)| < \infty$. But

$$\limsup_{n\to\infty} \mu_n(\psi) \leq \sup \psi < \infty$$

by Lemma 4.1.5 and

$$\begin{aligned} \liminf_{n\to\infty} \mu_n(\psi) &= \liminf_{n\to\infty} \frac{p((1+t_n)\psi + \phi_n) - \mu_n(\phi_n) - h(\mu_n)}{1+t_n} \\ &\geq p(\psi) - \sup_{\nu \in \mathcal{M}(\mathcal{T})} h(\nu) \qquad \text{by (4.5)} \\ &\geq p(\psi) - p(0) > -\infty . \end{aligned}$$

Therefore $h(\mu) + \mu(\psi) \geq p(\psi)$, i.e., $\mu \in ES(\psi, \mathcal{T})$.

b) If $(\mu_n)_n$ does not converge to μ, then there is a subsequence $(\mu_{n_i})_i$ converging to some $\nu \in \mathcal{M}(\mathcal{T}) \setminus \{\mu\}$, because $\mathcal{M}(\mathcal{T})$ is weakly compact. By the first part, $\nu \in ES(\psi, \mathcal{T}) = \{\mu\}$, a contradiction. □

4.3 Equilibrium states and convex geometry

We proceed to a geometric characterization of equilibrium states. Recall that $C^*(X)$ denotes the space of real-valued continuous linear functionals on $C(X)$ and that each such functional can be represented as the difference of two mutually singular finite Borel measures on X. In particular, each positive functional $\nu \in C^*(X)$ with the property $\nu(1) = 1$ determines a Borel probability measure on X (see A.4.31). In this sense $\mathcal{M}(\mathcal{T}) \subseteq C^*(X)$, and we extend the notation $(T\nu)(\psi) = \nu(\psi \circ T)$ from measures ν to general $\nu \in C^*(X)$.

4.3.1 Definition *a) $\nu \in C^*(X)$ is a tangent functional of the pressure function p at the point $\psi \in USC(X)$ if $p(\psi + \phi) \geq p(\psi) + \nu(\phi)$ for all $\phi \in C(X)$. The set of tangent functionals of p at ψ is denoted by $\mathcal{D}_\psi(p)$.*

b) If $\mathcal{D}_\psi(p) = \{\nu\}$, then p is called differentiable at $\psi \in USC(X)$ with derivative ν.

4.3.2 Remarks

a) In a broader context the notion of a tangent functional is discussed in [16, V.9].

b) Before we justify this definition of differentiability in Theorem 4.3.5, we show in the next theorem that for $\psi \in C(X)$ the tangent functionals of p at ψ are just the equilibrium states for ψ. In fact, $ES(\psi, \mathcal{T}) \subseteq \mathcal{D}_\psi(p)$ is true without any further assumption even for each $\psi \in USC(X)$, because if $\mu \in ES(\psi, \mathcal{T})$, then $p(\psi) = h(\mu) + \mu(\psi)$ and hence

$$p(\psi + \phi) \geq h(\mu) + \mu(\psi + \phi) = p(\psi) + \mu(\phi) \quad \text{for all } \phi \in C(X)\,,$$

i.e., $\mu \in \mathcal{D}_\psi(p)$. ◇

4.3.3 Theorem (Equilibrium states as tangent functionals of the pressure) *Let $(X, \mathcal{T})$ be a t.d.s., let $\psi \in C(X)$, and suppose that the entropy function h is upper semicontinuous. Then $ES(\psi, \mathcal{T}) = \mathcal{D}_\psi(p)$.*

4.3.4 Remark Suppose that h is upper semicontinuous. As $ES(\psi, \mathcal{T}) \neq \emptyset$ by Theorem 4.2.3, the preceding theorem shows that $\mathcal{D}_\psi(p) \neq \emptyset$ and that p is differentiable at ψ if and only if ψ has a unique equilibrium state.

Theorem 4.2.11, too, can be reinterpreted in a geometric way: if p is differentiable at $\psi \in USC(X)$ and at $\psi + \phi$ for ϕ in a neighbourhood of the constant function $0 \in C(X)$, then the derivative of p at $\psi + \phi$ depends continuously on ϕ.

In the context of statistical mechanics, nonuniqueness of the equilibrium state, and hence nondifferentiability of the pressure function, is interpreted as a *phase transition.* ◇

Proof of the theorem: In view of Remark 4.3.2 it remains to show that $\mathcal{D}_\psi(p) \subseteq ES(\psi, \mathcal{T})$. Observe that $-\infty < \inf \psi \leq p(\psi) \leq p(0) + \sup \psi < \infty$ since $p(0) = \sup_\nu h(\nu) < \infty$ by the upper semicontinuity of h, see Theorem 4.2.3. Suppose now that $\nu \in \mathcal{D}_\psi(p)$. Then

$$p(\psi) + 1 = p(\psi + 1) \geq p(\psi) + \nu(1) \quad \text{and} \quad p(\psi) - 1 = p(\psi - 1) \geq p(\psi) - \nu(1)$$

imply $\nu(1) = 1$. As the pressure function is isotonic, for each $\phi \geq 0$

$$p(\psi) \geq p(\psi - \phi) \geq p(\psi) - \nu(\phi), \text{ i.e., } \nu(\phi) \geq 0 .$$

Therefore ν represents a Borel probability measure on X. Furthermore, because of the tangent property of ν and Theorem 4.1.10, for each $T \in \mathcal{T}$, $\phi \in C(X)$, and $t \in \mathbb{R}$

$$t\,\nu(\phi \circ T - \phi) \leq p(\psi + t\phi \circ T - t\phi) - p(\psi) = 0 .$$

As t can be positive or negative, it follows that $\nu(\phi \circ T) = \nu(\phi)$ for all $T \in \mathcal{T}$ and $\phi \in C(X)$, i.e., ν is $\mathcal{T}$-invariant. To prove that ν is an equilibrium state we use the characterization of the semicontinuity of h given in Theorem 4.2.9 and conclude that

$$\begin{aligned} h(\nu) &= \inf_{\phi \in C(X)} [p(\psi + \phi) - \nu(\psi + \phi)] \quad \text{(observe that } \psi \in C(X)) \\ &\geq \inf_{\phi \in C(X)} [p(\psi) + \nu(\phi) - \nu(\psi + \phi)] \\ &= p(\psi) - \nu(\psi) \\ &\geq h(\nu) . \end{aligned}$$

It follows that $\nu \in ES(\psi, \mathcal{T})$. □

4.3.5 Theorem (Uniqueness of equilibr. states and Gateaux differentiability) *Suppose that h is upper semicontinuous. Then p is differentiable at $\psi \in C(X)$ with derivative μ if and only if*

$$\lim_{t \to 0} \frac{p(\psi + t\,\phi) - p(\psi)}{t} = \mu(\phi) \tag{4.6}$$

for each $\phi \in C(X)$. (We thus see that differentiability in our sense is just Gateaux differentiability, see e.g., [24].)

Proof: Suppose first that p is differentiable at ψ with derivative μ, i.e., that $ES(\psi, \mathcal{T}) = \mathcal{D}_\psi(p) = \{\mu\}$. For $t \in \mathbb{R}$ and $\phi \in C(X)$ pick some $\mu_t \in ES(\psi + t\phi)$. Then $\lim_{t \to 0} \mu_t = \mu$ by Theorem 4.2.11. As $\mu \in \mathcal{D}_\psi(p)$ we have

$$t\,\mu(\phi) \leq p(\psi + t\,\phi) - p(\psi) ,$$

and as $\mu_t \in \mathcal{D}_{\psi + t\phi}(p)$ we have

$$-t\,\mu_t(\phi) \leq p((\psi + t\,\phi) - t\,\phi) - p(\psi + t\,\phi) = -(p(\psi + t\,\phi) - p(\psi)) .$$

Hence, if $t > 0$,

$$\mu(\phi) \leq \frac{p(\psi + t\,\phi) - p(\psi)}{t} \leq \mu_t(\phi)\,,$$

otherwise the inequalities are reversed. In any case it follows that

$$\lim_{t\to 0} \frac{p(\psi + t\,\phi) - p(\psi)}{t} = \mu(\phi)$$

because $\lim_{t\to 0} \mu_t(\phi) = \mu(\phi)$.

Suppose now that (4.6) holds for all $\phi \in C(X)$. If ν is an arbitrary element in $\mathcal{D}_\psi(p)$ (which is nonempty by Theorem 4.2.3), then

$$\mu(\phi) = \lim_{t\downarrow 0} \frac{p(\psi + t\,\phi) - p(\psi)}{t} \geq \nu(\phi)$$

and similarly

$$\mu(\phi) = \lim_{t\uparrow 0} \frac{p(\psi + t\,\phi) - p(\psi)}{t} \leq \nu(\phi)\,.$$

Hence $\nu(\phi) = \mu(\phi)$ for all $\phi \in C(X)$ and all $\nu \in \mathcal{D}_\psi(p)$, i.e., $\mathcal{D}_\psi(p) = \{\mu\}$. □

In terms of tangent functionals the convexity of the pressure function is expressed by

4.3.6 Corollary *Let $\psi, \phi \in C(X)$ and consider $\mu_i \in ES(\psi + t_i\phi, \mathcal{T})$, $i = 1, 2$, where $t_1 < t_2$ are any reals. Then $\mu_1(\phi) \leq \mu_2(\phi)$.*

Proof: In view of Theorem 4.3.3,

$$(t_2 - t_1)\,\mu_1(\phi) \leq p(\psi + t_2\phi) - p(\psi + t_1\phi) \leq -(t_1 - t_2)\,\mu_2(\phi)\,.$$

As $t_1 < t_2$ the claim follows. □

We finish this section with a result extending the affinity of the entropy function to general ergodic decompositions.

4.3.7 Theorem (Affinity of the entropy function)
Let $(X, \mathcal{T})$ be a t.d.s. and suppose that the entropy function $h : \mathcal{M}(\mathcal{T}) \to \mathbb{R}$ is upper semicontinuous (with respect to the weak topology on $\mathcal{M}(\mathcal{T})$). Then each $\mu \in \mathcal{M}(\mathcal{T})$ satisfies

$$h(\mu) = \int h(\mu_x)\,d\mu(x)\,,$$

where $\mu = \int \mu_x\,d\mu(x)$ is the ergodic decomposition of μ.

Proof: Let $\mathcal{B}$ be the Borel σ-algebra of X.

Step 1: The map $x \mapsto h(\mu_x)$ is $\mathcal{B}$-measurable, because $C(X)$ contains a countable dense subset $\{\psi_k : k \in \mathbb{N}\}$ such that in view of Theorem 4.2.9

$$h(\mu_x) = \inf_{\psi \in C(X)} (p(\psi) - \mu_x(\psi)) = \inf_{k \geq 0} (p(\psi_k) - \mu_x(\psi_k))$$

where the $x \mapsto \mu_x(\psi_k)$ are measurable (see A.4.42).

Step 2: By Step 1, $\int h(\mu_x)\, d\mu(x)$ is well defined and in view of Theorem 4.2.9

$$\begin{aligned} \int h(\mu_x)\, d\mu(x) &= \int \inf_{\psi \in C(X)} (p(\psi) - \mu_x(\psi))\, d\mu(x) \\ &\leq \inf_{\psi \in C(X)} \int (p(\psi) - \mu_x(\psi))\, d\mu(x) = \inf_{\psi \in C(X)} (p(\psi) - \mu(\psi)) = h(\mu)\,. \end{aligned}$$

This proves the "$\geq$"-direction of the theorem.

Step 3: We decompose the space X into a finite number of $\mathcal{T}$-invariant measurable subsets on which $h(\mu_x)$ is nearly constant. To this end let ρ be a metric for the topology on X and set

$$L_n := \{\psi \in C(X) : |\psi(u)| \leq n \text{ and } |\psi(u) - \psi(v)| \leq n \cdot \rho(u, v)\ \forall u, v \in X\}\,.$$

As the L_n are families of equicontinuous functions, they are relatively compact by the theorem of Arzelà and Ascoli, and as the sets L_n are obviously closed in $C(X)$, they are in fact compact. Furthermore, the set $\bigcup_{n>0} L_n$ is dense in $C(X)$ since it is the set of all Lipschitz-continuous functions on X (see Appendix, A.1).

Let $\ell_n(\mu) := \inf_{\psi \in L_n} (p(\psi) - \mu(\psi))$. Then

$$\infty > p(0) \geq \ell_n(\mu) \searrow \inf_{\psi \in C(X)} (p(\psi) - \mu(\psi)) = h(\mu) \geq 0$$

for each $\mu \in \mathcal{M}(\mathcal{T})$, and by the same argument as in Step 1, the maps $x \mapsto \ell_n(\mu_x)$ are measurable. Hence $\int h(\mu_x)\, d\mu(x) = \lim_{n \to \infty} \int \ell_n(\mu_x)\, d\mu(x)$. Fix $\epsilon > 0$. There is $N > 0$ such that

$$\int h(\mu_x)\, d\mu(x) \geq \int \ell_N(\mu_x)\, d\mu(x) - \epsilon\,.$$

As L_N is compact, there are $\psi_1, \ldots, \psi_k \in L_N$ such that for all $\psi \in L_N$ there exists $i \in \{1, \ldots, k\}$ with $\|\psi - \psi_i\|_\infty < \epsilon$. As the pressure function p and all the μ_x are Lipschitz-continuous with Lipschitz constant 1 as functions from $C(X)$ to $\mathbb{R}$ (for p see Theorem 4.1.10, for the μ_x this is trivial), it follows that

$$\ell_N(\mu_x) \geq \min_{i=1,\ldots,k} (p(\psi_i) - \mu_x(\psi_i)) - 2\epsilon\,. \tag{4.7}$$

Define $\kappa : X \to \{1, \ldots, k\}$ by

$$\kappa(x) := \min\{i \in \{1, \ldots, k\} : \ell_N(\mu_x) \geq p(\psi_i) - \mu_x(\psi_i) - 2\epsilon\}\,.$$

Because of (4.7) the map κ is well defined and, as $\mu_{Tx} = \mu_x$ for all $x \in X$ and $T \in \mathcal{T}$ (see Theorem 2.3.3), it is $\mathcal{T}$-invariant. Hence also the sets $A_i := \{x \in X : \kappa(x) = i\}$ are $\mathcal{T}$-invariant. We may assume that $\mu(A_i) > 0$ for all i. Denote $\mu_i := \mu(\,.\mid A_i)$. Then the measures μ_i are $\mathcal{T}$-invariant and

$$\begin{aligned}\int_{A_i} \ell_N(\mu_x)\,d\mu(x) &\geq \int_{A_i} (p(\psi_i) - \mu_x(\psi_i))\,d\mu(x) - 2\epsilon\mu(A_i)\\ &= \mu(A_i)\cdot(p(\psi_i) - \mu_i(\psi_i) - 2\epsilon)\\ &\geq \mu(A_i)\cdot(h(\mu_i) - 2\epsilon)\,.\end{aligned}$$

Now Theorem 3.3.2, which asserts the convex affinity of the entropy function, implies

$$\begin{aligned}\int h(\mu_x)\,d\mu(x) &\geq \sum_{i=1}^{k}\int_{A_i} \ell_N(\mu_x)\,d\mu(x) - \epsilon\\ &\geq \sum_{i=1}^{k} \mu(A_i)h(\mu_i) - 3\epsilon = h(\mu) - 3\epsilon\,.\end{aligned}$$

Taking the limit $\epsilon \to 0$ this finishes the proof of the theorem. □

4.3.8 Remark This theorem remains true, if one drops the upper semicontinuity assumption on h, see [60, Theorem 8.4]. ◇

4.3.9 Theorem (Ergodic decomposition of equilibrium states)
Under the same assumptions as in Theorem 4.3.7 let μ be an equilibrium state for $\psi \in USC(X)$ with ergodic decomposition $\mu = \int \mu_x\,d\mu(x)$. Then μ-almost all μ_x are equilibrium states for ψ.

Proof: Because of Theorem 4.3.7

$$\begin{aligned}p(\psi) &= h(\mu) + \mu(\psi) = \int h(\mu_x)\,d\mu(x) + \int \mu_x(\psi)\,d\mu(x)\\ &= \int (h(\mu_x) + \mu_x(\psi))d\mu(x)\,,\end{aligned}$$

and as $h(\mu_x) + \mu_x(\psi) \leq p(\psi)$ for all x, the equality $h(\mu_x) + \mu_x(\psi) = p(\psi)$ follows for μ-a.e. x. □

4.3.10 Exercise Let $(X, \mathcal{T})$ be a t.d.s. with upper semicontinuous entropy function. Denote by $\mathcal{B}$ the Borel σ-algebra of X. Prove that the set $\mathcal{K}$ of finite convex combinations of ergodic $\mathcal{T}$-invariant measures is dense in $\mathcal{M}(\mathcal{T})$. *Hint:* Show that otherwise there would be $\psi \in C(X)$ and $\mu \in ES(\psi, \mathcal{T})$ such that $\mu \notin \bar{\mathcal{K}}$. ◇

4.4 The variational principle

The goal of this section is to provide a more explicit expression for the pressure $p(\psi)$ (an identity that is often referred to as the *variational principle*) and to describe a scheme to approximate equilibrium states by purely atomic measures. We follow essentially Misiurewicz's proof of the variational principle [41], but organize it in such a way that it will allow us in Section 5.3 to identify equilibrium states in certain circumstances as Gibbs measures, a notion that generalizes the like concept for finite systems from Section 1.1. Because of the central role of the variational principle for the theory of equilibrium states we mention also Walters' important publication [61].

Let $\mathcal{T}$ be a continuous action of $G = \mathbb{Z}^d$ or $G = \mathbb{Z}^d_+$ on the compact metric space (X, ρ), and recall the notations $\Lambda_n = \{g \in G : |g_i| < n \ \forall i\}$ and $\lambda_n = |\Lambda_n|$. For $\delta > 0$, $n \in \mathbb{N}$ and $x \in X$ let

$$U(n, \delta, x) := \{y \in X : \rho(T^g y, T^g x) < \delta \ \forall g \in \Lambda_n\} = \bigcap_{g \in \Lambda_n} T^{-g} B_\delta(T^g x)$$

where $B_\delta(x)$ is the open δ-ball around x. As all T^g are continuous, $U(n, \delta, x)$ is an open neighbourhood of x.

4.4.1 Definition *A subset E of X is called*

▷ *(n, δ)-separated, if $y \notin U(n, \delta, x)$ for all $x, y \in E$ with $x \neq y$,*

▷ *maximal (n, δ)-separated, if E is (n, δ)-separated and if there is no (n, δ)-separated set that properly contains E,*

▷ *(n, δ)-dense, if $X = \bigcup_{x \in E} U(n, \delta, x)$.*

The following two facts are obvious from these definitions:

▷ A maximal (n, δ)-separated set is (n, δ)-dense.

▷ As X is compact, (n, δ)-separated sets are finite, and for each $\delta > 0$ and $n \in \mathbb{N}$ there are maximal (n, δ)-separated sets.

4.4.2 Example (((n, δ)-separated sets in shift spaces) Recall from Example 4.2.2 that $X = \Omega = \Sigma^G$ is endowed with the metric $d(\omega, \omega') = 2^{-n(\omega, \omega')}$ where $n(\omega, \omega') = \sup\{n \in \mathbb{N} : \omega_g = \omega'_g \ \forall g \in \Lambda_n\}$. Then $U(n, 2^{-k}, \omega) = \{\omega' \in \Omega : \omega'_g = \omega_g \ \forall g \in \Lambda_{n+k}\}$, whence $U(n, 2^{-(k+j)}, \omega) = U(n + j, 2^{-k}, \omega)$ for all $j, k, n \in \mathbb{N}$ and all $\omega \in \Omega$. In particular, a set $E \subseteq \Omega$ is [maximal] $(n, 2^{-(j+k)})$-separated if and only if it is [maximal] $(n+k, 2^{-j})$-separated ($j, k \in \mathbb{N}$). Similarly, E is maximal $(n, 2^{-k})$-separated ($k \in \mathbb{N}$) if and only if for each $v \in \Sigma^{\Lambda_{n+k}}$ there is exactly one $\omega \in E$ such that $\omega_g = v_g$ for all $g \in \Lambda_{n+k}$. ◇

For $\psi \in USC(X)$, $n \in \mathbb{N}$, $\delta > 0$ and a finite $E \subset X$ let $\psi_n := \sum_{g \in \Lambda_n} \psi \circ T^g$ and

$$\begin{aligned} P(\psi, E) &:= \log \sum_{x \in E} \exp \psi(x) , \\ P_{n,\delta}(\psi) &:= \sup\{P(\psi_n, E) : E \text{ is } (n,\delta)\text{-separated}\} , \\ &= \sup\{P(\psi_n, E) : E \text{ is maximal } (n,\delta)\text{-separated}\} \quad \text{and} \\ p_\delta(\psi) &:= \limsup_{n\to\infty} \frac{1}{\lambda_n} P_{n,\delta}(\psi) . \end{aligned}$$

It follows immediately from these definitions that $P_{n,\delta}(\psi)$ and $p_\delta(\psi)$ increase if δ decreases. The main result of this section (Theorem 4.4.11) is that $p(\psi) = \sup_{\delta>0} p_\delta(\psi)$ for all $\psi \in USC(X)$.

4.4.3 Example (Pressure in shift spaces I) (Continuation of Example 4.4.2) For shift spaces $X = \Omega = \Sigma^G$ and $\psi \in USC(X)$ with $\inf \psi > -\infty$

$$\begin{aligned} & P_{n,2^{-k}}(\psi) \\ &= \sup\{P(\psi_n, E) : E \text{ is } (n+k,1)\text{-separated}\} \\ &\le \sup\{P(\psi_{n+k}, E) + (\lambda_{n+k} - \lambda_n) \cdot |\inf \psi| : E \text{ is } (n+k,1)\text{-separated}\} \\ &= P_{n+k,1}(\psi) + (\lambda_{n+k} - \lambda_n) \cdot |\inf \psi| . \end{aligned}$$

Hence

$$\begin{aligned} p_{2^{-k}}(\psi) &= \limsup_{n\to\infty} \frac{1}{\lambda_n} P_{n,2^{-k}}(\psi) \\ &\le \limsup_{n\to\infty} \frac{\lambda_{n+k}}{\lambda_n} \cdot \frac{1}{\lambda_{n+k}} P_{n+k,1}(\psi) + \limsup_{n\to\infty} \frac{\lambda_{n+k} - \lambda_n}{\lambda_n} \cdot |\inf \psi| \\ &= p_1(\psi) , \end{aligned}$$

as $\lim_{n\to\infty} \frac{\lambda_{n+k}}{\lambda_n} = 1$. Taking into account the monotonicity of $p_\delta(\psi)$ as a function of δ it follows that

$$p_\delta(\psi) = p_1(\psi) \quad \text{for all } \delta \in (0,1).$$

If $\inf \psi = -\infty$, the same estimate still works under the additional hypothesis that $T^{-g}\{\psi = -\infty\} \subseteq \{\psi = -\infty\}$ for all $g \in G$ and that $\underline{\psi} := \inf\{\psi(\omega) : \psi(\omega) > -\infty\} > -\infty$. In this case $\psi(\omega) = -\infty$ if $\psi(T^g\omega) = -\infty$ for some $g \in G$, and it follows that $P(\psi_n, E) \le P(\psi_{n+k}, E) + (\lambda_{n+k} - \lambda_n) \cdot |\underline{\psi}|$ since $P(\psi_n, E) = P(\psi_n, E \cap \{\psi > -\infty\})$ for all $n \ge 1$. The functions ψ^* from Remark 4.2.7 are typical examples for this situation. ◇

This leads to the following definition:

4.4.4 Definition

$$USC_0(X) := \{\psi \in USC(X) : \{\psi = -\infty\} \textit{ is } \mathcal{T}\textit{-invariant and } \underline{\psi} > -\infty\} .$$

4.4.5 Example (Pressure in shift spaces II) (Continuation of Example 4.4.3) For shift spaces and functions $\psi \in USC_0(\Omega)$ it is in fact relatively easy to establish the inequality $p(\psi) \leq p_1(\psi)$. We give the proof below. For more general dynamical systems the corresponding estimate is proved in Theorem 4.4.11, and the reverse inequality is derived in Corollary 4.4.9.

Denote by α the partition of X into cylinder sets $[\sigma]_0, \sigma \in \Sigma$. Then α is a finite generator for the shift action $\mathcal{T}$ on X (Example 3.2.20) so that $h(\mu) = h(\mu, \alpha, \mathcal{T})$ for all $\mu \in \mathcal{M}(\mathcal{T})$ by Theorem 3.2.18. We have to show that

$$h(\mu, \alpha, \mathcal{T}) + \mu(\psi) \leq p_1(\psi) \quad \text{for all } \mu \in \mathcal{M}(\mathcal{T}). \tag{4.8}$$

To this end let $\beta_n := \{A \in \alpha^{\Lambda_n} : \mu(A) > 0\}$. For each $B \in \beta_n$ there exists a point $x_B \in B$ such that

$$\int_B \psi_n \, d\mu \leq \psi_n(x_B) \cdot \mu(B) .$$

Let $E := \{x_B : B \in \beta_n\}$. By the elementary version of Jensen's inequality

$$\begin{aligned} & H_\mu(\alpha^{\Lambda_n}) + \lambda_n \cdot \mu(\psi) \qquad\qquad (4.9) \\ = \; & H_\mu(\beta_n) + \mu(\psi_n) \leq \sum_{B \in \beta_n} \mu(B) \log \frac{\exp \psi_n(x_B)}{\mu(B)} \\ \leq \; & \log \left(\sum_{B \in \beta_n} \mu(B) \frac{\exp \psi_n(x_B)}{\mu(B)} \right) = \log \left(\sum_{x \in E} \exp \psi_n(x) \right) \\ = \; & P(\psi_n, E) . \end{aligned}$$

Because of Example 4.4.2 the set E is $(n, 1)$-separated. Therefore

$$\frac{1}{\lambda_n} H_\mu(\alpha^{\Lambda_n}) + \mu(\psi) \leq \frac{1}{\lambda_n} P(\psi_n, E) \leq \frac{1}{\lambda_n} P_{n,1}(\psi)$$

so that

$$h(\mu, \alpha, \mathcal{T}) + \mu(\psi) = \lim_{n \to \infty} \frac{1}{\lambda_n} H_\mu(\alpha^{\Lambda_n}) + \mu(\psi) \leq \limsup_{n \to \infty} \frac{1}{\lambda_n} P_{n,1}(\psi) = p_1(\psi) ,$$

i.e., (4.8). ◇

4.4.6 Exercise Let $X = \Omega = \Sigma^G$ be a shift space and consider a local energy function ψ that depends only on the zero coordinate, $\psi(\omega) = u(\omega_0)$ for some function $u : \Sigma \to \mathbb{R}$. Prove that $p_1(\psi) = p(\psi)$ (see Example 4.2.2). ◇

4.4.7 Exercise Let $X = \Omega = \Sigma^{\mathbb{Z}}$ be a one-dimensional shift space. Consider a local energy function ψ that depends only on two coordinates, $\psi(\omega) = f(\omega_0, \omega_1)$. Denote by A the $\Sigma \times \Sigma$ matrix with entries $a_{\sigma\tau} = e^{f(\sigma,\tau)}$. Use the theorem of

Frobenius and Perron (A.3.1) to prove that $p_1(\psi)$ is the logarithm of the largest eigenvalue of the matrix A.

Comment: It is shown in A.3.2 that the unique normalized right eigenvector v for the largest eigenvalue r of A can be used to define a stochastic matrix $Q = r^{-1}\,\mathrm{diag}(v)^{-1}A\,\mathrm{diag}(v)$. Define $\tilde{f} : \Sigma^2 \to \mathbb{R}$, $\tilde{f}(\sigma,\tau) := \log q_{\sigma\tau} = \log a_{\sigma\tau} + \log v_\tau - \log v_\sigma - \log r$. $\tilde{f}$ defines the local energy function $\tilde{\psi} := e^{\tilde{f}}$. Let $\phi(\omega) := v_{\omega_0}$. Then $\tilde{\psi} = \psi + \phi \circ T - \phi - \log r$, and $p_1(\tilde{\psi})$ is the logarithm of the largest eigenvalue of Q, i.e., $p_1(\tilde{\psi}) = 0 = p_1(\psi) - \log r$. Compare this to the identity $p(\tilde{\psi}) = p(\psi) - \log r$ which follows from Theorem 4.1.10. ◇

Now we turn to the construction of equilibrium states $\mu \in ES(\psi, \mathcal{T})$ for functions $\psi \in USC(X)$. With a view to later applications we want to keep the construction as flexible as possible. Therefore we consider sequences of functions $\psi^{(n)} \in USC(X)$ that approximate the functions ψ_n in the sense that

$$\|\psi^{(n)} - \psi_n\|_\infty = o(\lambda_n)\,. \tag{4.10}$$

For a first reading the reader may just replace $\psi^{(n)}$ by ψ_n.

As functions in $USC(X)$ are allowed to take the value $-\infty$, the difference $(\psi^{(n)} - \psi_n)$ must be interpreted carefully. We think of it as a bounded function $\Delta^{(n)}$ such that $\psi^{(n)} = \psi_n + \Delta^{(n)}$ and $\|\Delta^{(n)}\|_\infty = o(\lambda_n)$ which is well defined even if $\psi^{(n)}$ and ψ_n take the value $-\infty$.

Fix $\delta > 0$. We perform our construction with any sequence $(E_n)_{n\in\mathbb{N}}$ of (n,δ)-separated subsets of X. Observe that

$$P(\psi_n, E_n) \le P(\psi^{(n)}, E_n) + \|\psi^{(n)} - \psi_n\|_\infty\,. \tag{4.11}$$

The sets E_n support probability measures π_n defined by

$$\pi_n = \sum_{z\in E_n} \delta_z \cdot \frac{\exp \psi^{(n)}(z)}{\exp P(\psi^{(n)}, E_n)} \tag{4.12}$$

where the $\psi^{(n)}$ are as described above. (Indeed, the π_n are probability measures because $\sum_{z\in E_n} \exp \psi^{(n)}(z) = \exp P(\psi^{(n)}, E_n)$. They are elementary Gibbs measures on E_n in the sense of Section 1.1.)

As the weak limit points of the sequence $(\pi_n)_{n\in\mathbb{N}}$ are in general not $\mathcal{T}$-invariant, we take averages and consider probability measures

$$\mu_n = \frac{1}{\lambda_n} \sum_{g\in\Lambda_n} T^g \pi_n\,.$$

Observe that

$$\mu_n(\phi) = \frac{1}{\lambda_n}\,\pi_n(\phi_n) \tag{4.13}$$

for all $\phi \in USC(X)$. Let (n_i) be any increasing sequence of integers such that

$$\lim_{i\to\infty} \frac{1}{\lambda_{n_i}} P(\psi_{n_i}, E_{n_i}) = \limsup_{n\to\infty} \frac{1}{\lambda_n} P(\psi_n, E_n) \tag{4.14}$$

and

$$\lim_{i\to\infty} \mu_{n_i} =: \mu \quad \text{exists.}$$

Such a sequence exists because of the weak compactness of the set of all Borel probabilities on X. The (nonempty!) set of all limit measures μ that we can obtain from this construction is denoted by $\mathcal{M}((E_n)_n, \delta, \psi, \mathcal{T})$.

If we want to come close to constructing equilibrium states in this way, we need to assume that the sets E_{n_i} in (4.14) are such that

$$P(\psi_{n_i}, E_{n_i}) \geq P_{n_i,\delta}(\psi) - o(\lambda_{n_i}). \tag{4.15}$$

Such sets exist by definition of $P_{n,\delta}(\psi)$. The (nonempty!) set of limiting measures one can obtain from such sets E_n is denoted by $CES(\delta, \psi, \mathcal{T})$. This abbreviation stands for *constructible δ-equilibrium state* and is partially justified by the following lemma. In fact, we will see later that if $p_\delta(\psi) = p(\psi)$ for some $\delta > 0$, then the closed convex hull of $CES(\delta, \psi, \mathcal{T})$ coincides with $ES(\psi, \mathcal{T})$. Otherwise, constructible δ-equilibrium states are not necessarily equilibrium states.

4.4.8 Lemma *Let $\psi \in USC(X)$, $\delta > 0$, and suppose that there are a partition α and a sequence of integers $(M_n)_{n\in\mathbb{N}}$ such that $|E_n \cap A| \leq M_n$ for each $A \in \alpha^{\Lambda_n}$ and $\log M_n = o(\lambda_n)$. Let $\mu \in \mathcal{M}((E_n)_n, \delta, \psi, \mathcal{T})$ and suppose that*

$$\lim_{m\to\infty} \frac{1}{\lambda_m} \left(\sum_{A\in\alpha^{\Lambda_m}} \mu(\bar{A} \setminus A) \,|\log \mu(A)| + \log a_m \right) = 0 \tag{4.16}$$

where $\bar{A}$ is the topological closure of A and $a_m = |\{A \in \alpha^{\Lambda_m} : \mu(\bar{A}) > e^{-1}\}| + 1$, see Theorem 4.2.4. Then μ is $\mathcal{T}$-invariant and satisfies

$$h(\mu, \alpha, \mathcal{T}) + \mu(\psi) \geq \limsup_{n\to\infty} \frac{1}{\lambda_n} P(\psi_n, E_n). \tag{4.17}$$

Proof: Let μ be in $\mathcal{M}((E_n)_n, \delta, \psi, \mathcal{T})$, $\mu = \lim_{i\to\infty} \mu_{n_i}$. The $\mathcal{T}$-invariance of μ follows from Lemma 4.1.2.

We begin the proof of (4.17) by showing that

$$P(\psi^{(n)}, E_n) \leq H_{\pi_n}(\alpha^{\Lambda_n}) + \pi_n(\psi^{(n)}) + \log M_n. \tag{4.18}$$

For each $A \in \alpha^{\Lambda_n}$ we have

$$\begin{aligned} &-\pi_n(A) \log \pi_n(A) + \sum_{z\in E_n\cap A} \pi_n\{z\} \log \pi_n\{z\} \\ &= \pi_n(A) \sum_{z\in E_n\cap A} \frac{\pi_n\{z\}}{\pi_n(A)} \log \frac{\pi_n\{z\}}{\pi_n(A)} \\ &\geq -\pi_n(A) \cdot \log |E_n \cap A| \geq -\pi_n(A) \cdot \log M_n. \end{aligned}$$

Summing this inequality over all $A \in \alpha^{\Lambda_n}$ yields

$$H_{\pi_n}(\alpha^{\Lambda_n}) + \sum_{z \in E_n} \pi_n\{z\} \log \pi_n\{z\} \geq -\log M_n$$

so that

$$\begin{aligned} H_{\pi_n}(\alpha^{\Lambda_n}) + \pi_n(\psi^{(n)}) &\geq \sum_{z \in E_n} \pi_n\{z\} \underbrace{(\psi^{(n)}(z) - \log \pi_n\{z\})}_{=P(\psi^{(n)},E_n)} - \log M_n \\ &= P(\psi^{(n)}, E_n) - \log M_n \,, \end{aligned}$$

that is (4.18).

Fix $m \in \mathbb{N}$, $m > 0$, and consider $n \geq 2m$. We are going to pave the grid Λ_n except for a small number of points with copies of Λ_m. This will be done in various ways such that the exceptional set of points is different in each variation of the construction. Denote by l_m the side length of Λ_m. For $g \in \Lambda_m$ let

$$S(g) := \{g + l_m h : h \in G, g + l_m h \in \Lambda_{n-m}\} \,.$$

$S(g)$ is a subgrid of Λ_{n-m} with mesh width m. Let $R(g) := \Lambda_n \setminus (S(g) + \Lambda_m)$. (For $G = \mathbb{Z}_+^2$, $m = l_m = 2$, $n = 9$ and $g = (2,1)$ the situation is sketched in Figure 4.1.) It is easy to see that

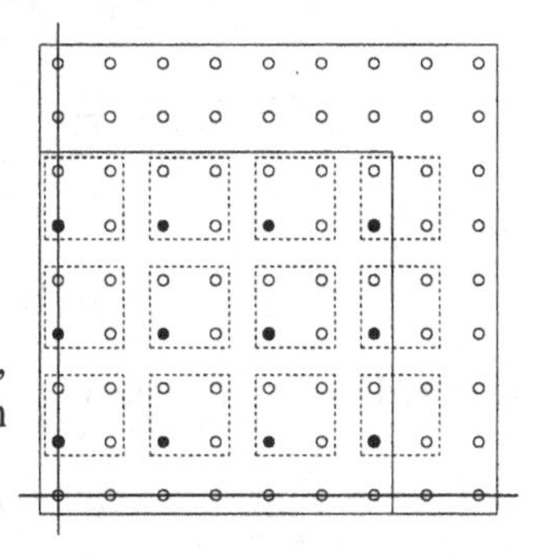

Figure 4.1: Sketch of the sets Λ_n (large solid box), Λ_{n-m} (smaller solid box), $S(g)$ (black dots) and $S(g) + \Lambda_m$ (dashed boxes).

$$|S(g)| \geq \left(\frac{l_{n-2m}}{l_m}\right)^d = \frac{\lambda_{n-2m}}{\lambda_m} \,,$$

$$\begin{aligned} |R(g)| &= \lambda_n - |S(g)| \cdot \lambda_m \\ &\leq \lambda_n - \lambda_{n-2m} =: \gamma_{m,n} \,, \end{aligned}$$

and

$$\alpha^{\Lambda_n} = \left(\bigvee_{s \in S(g)} T^{-s} \alpha^{\Lambda_m} \right) \vee \bigvee_{k \in R(g)} T^{-k} \alpha \,.$$

Therefore each $g \in \Lambda_m$ admits the estimate

$$\begin{aligned} P(\psi^{(n)}, E_n) &\leq H_{\pi_n}(\alpha^{\Lambda_n}) + \pi_n(\psi^{(n)}) + \log M_n \\ &\leq \sum_{s \in S(g)} H_{\pi_n}(T^{-s}\alpha^{\Lambda_m}) + \pi_n(\psi^{(n)}) + \sum_{k \in R(g)} H_{\pi_n}(T^{-k}\alpha) + o(\lambda_n) \\ &\leq \sum_{s \in S(g)} H_{\pi_n}(T^{-s}\alpha^{\Lambda_m}) + \pi_n(\psi_n) + o(\lambda_n) + \gamma_{m,n} \log |\alpha| \end{aligned}$$

where we used (4.18) for the first inequality, Theorem 3.1.8-c for the second one and (4.10) for the third one. Summing up this estimate over all $g \in \Lambda_m$ we obtain

$$\begin{aligned}
&\lambda_m \cdot P(\psi^{(n)}, E_n) \\
&\leq \sum_{g\in\Lambda_m}\sum_{s\in S(g)} H_{\pi_n}(T^{-s}\alpha^{\Lambda_m}) + \lambda_m\left(\pi_n(\psi_n) + \gamma_{m,n}\log|\alpha| + o(\lambda_n)\right) \\
&\leq \sum_{k\in\Lambda_n} H_{\pi_n}(T^{-k}\alpha^{\Lambda_m}) + \lambda_m\left(\pi_n(\psi_n) + \gamma_{m,n}\log|\alpha| + o(\lambda_n)\right) . \qquad (4.19)
\end{aligned}$$

As the function $\mu \mapsto H_\mu(\beta)$ is concave for fixed β (see Theorem 3.3.2) we have

$$H_{\mu_n}(\alpha^{\Lambda_m}) \geq \frac{1}{\lambda_n}\sum_{k\in\Lambda_n} H_{T^k\pi_n}(\alpha^{\Lambda_m}) = \frac{1}{\lambda_n}\sum_{k\in\Lambda_n} H_{\pi_n}(T^{-k}\alpha^{\Lambda_m}) , \qquad (4.20)$$

and as $\pi_n(\psi_n) = \lambda_n \cdot \mu_n(\psi)$ by (4.13), we can combine (4.19) and (4.20) to obtain

$$\frac{1}{\lambda_n}\cdot P(\psi^{(n)}, E_n) \leq \frac{1}{\lambda_m} H_{\mu_n}(\alpha^{\Lambda_m}) + \mu_n(\psi) + \frac{\gamma_{m,n}}{\lambda_n}\log|\alpha| + \frac{o(\lambda_n)}{\lambda_n} . \qquad (4.21)$$

Since

$$\lim_{n\to\infty}\frac{\gamma_{m,n}}{\lambda_n} = \lim_{n\to\infty}\left(1 - \frac{\lambda_{n-2m}}{\lambda_n}\right) = 0$$

for fixed $m > 0$, this yields

$$\lim_{i\to\infty}\frac{1}{\lambda_{n_i}} P(\psi^{(n_i)}, E_{n_i}) \leq \frac{1}{\lambda_m}\limsup_{i\to\infty} H_{\mu_{n_i}}(\alpha^{\Lambda_m}) + \mu(\psi) . \qquad (4.22)$$

Observing (4.11), (4.14) and Theorem 4.2.4 it follows that

$$\begin{aligned}
\limsup_{n\to\infty}\frac{1}{\lambda_n} P(\psi_n, E_n) &\leq \limsup_{m\to\infty}\frac{1}{\lambda_m}\left(\limsup_{n_i\to\infty} H_{\mu_{n_i}}(\alpha^{\Lambda_m})\right) + \mu(\psi) \\
&\leq h(\mu, \alpha, \mathcal{T}) + \mu(\psi) .
\end{aligned}$$

□

4.4.9 Corollary *Let $\psi \in USC(X)$ and $\delta > 0$. Each $\mu \in CES(\delta, \psi, \mathcal{T})$ is $\mathcal{T}$-invariant and satisfies*

$$h(\mu) + \mu(\psi) \geq p_\delta(\psi) . \qquad (4.23)$$

Proof: We want to apply the preceding lemma to $\mu \in CES(\delta, \psi, \mathcal{T})$. So assume that $\mu \in \mathcal{M}((E_n)_n, \psi, \mathcal{T})$ for a sequence of (n, δ)-separated sets E_n that also satisfies (4.15). Then

$$p_\delta(\psi) = \limsup_{n\to\infty}\frac{1}{\lambda_n} P_{n,\delta}(\psi) \leq \limsup_{n\to\infty}\frac{1}{\lambda_n} P(\psi_n, E_n) \leq h(\mu) + \mu(\psi) ,$$

provided we find a finite partition α satisfying the assumptions of the preceding lemma. But the next lemma provides a finite partition α of X with $\operatorname{diam}(A) < \delta$ and $\mu(\partial A) = 0$ for all $A \in \alpha$. Then $y \in U(n, \delta, x)$ for all $x, y \in A \in \alpha^{\Lambda_n}$, and as E_n is (n, δ)-separated, it follows that $|A \cap E_n| \leq 1$ for all $A \in \alpha^{\Lambda_n}$. Condition (4.16) is trivially satisfied because $\mu(\partial A) = 0$ for all $A \in \alpha^{\Lambda_m}$ (Theorem 4.2.4). □

4.4.10 Lemma *Let μ be a Borel probability measure on a compact metric space (X, ρ) and let $\delta > 0$. There exists a finite partition α of X with* $\operatorname{diam}(A) < \delta$ *and $\mu(\partial A) = 0$ for all $A \in \alpha$.*

Proof: For $x \in X$ and $r > 0$ let $S_r(x) := \{y \in X : \rho(x, y) = r\}$. As $S_r(x) \cap S_{r'}(x) = \emptyset$ if $r \neq r'$, there are $r(x) \in (\frac{\delta}{4}, \frac{\delta}{2})$ such that $\mu(S_{r(x)}(x)) = 0$ for all $x \in X$. As the sets $B(x) := \{y \in X : \rho(x, y) < r(x)\}$ form an open cover of X, there are $x_1, \ldots, x_n$ such that $X = \bigcup_{i=1}^{n} B(x_i)$. Define inductively $A_1 := B(x_1)$ and $A_i := B(x_i) \setminus A_{i-1}$ for $i = 2, \ldots, n$. Then $\alpha := \{A_1, \ldots, A_n\}$ is a finite partition of X and for each A_i we have $\operatorname{diam}(A_i) \leq \operatorname{diam}(B(x_i)) \leq 2r(x_i) < \delta$ and $\partial A_i \subseteq \bigcup_{j=1}^{i} S_{r(x_j)}(x_j)$ so that $\mu(\partial A_i) = 0$. □

The central result of this section is

4.4.11 Theorem (Variational principle)
Let $\mathcal{T}$ be a continuous action of $G = \mathbb{Z}^d$ or $G = \mathbb{Z}_+^d$ on the compact metrizable space X and suppose that $\psi \in USC(X)$. Then

$$\sup_{\delta > 0} p_\delta(\psi) = \lim_{\delta \searrow 0} p_\delta(\psi) = p(\psi)$$

where $p(\psi) = \sup_{\mu \in \mathcal{M}(\mathcal{T})}(h(\mu) + \mu(\psi))$ was defined in (4.1).

4.4.12 Remark In the formulation of this theorem we did not refer to any particular metric on X, although the definition of the quantities p_δ depends on it. This is justified by the following observation: If $\tilde{\rho}$ is another metric on X equivalent to ρ (i.e., $\tilde{\rho}$ generates the same topology as ρ does), then, because of the compactness of X, for each $\delta > 0$ there is some $\tilde{\delta} > 0$ such that $\rho(x, y) < \delta$ if $\tilde{\rho}(x, y) < \tilde{\delta}$. Therefore a set $E \subseteq X$ which is (n, δ)-separated w.r.t. the metric ρ is $(n, \tilde{\delta})$-separated w.r.t. the metric $\tilde{\rho}$, so that $p_\delta(\psi) \leq \tilde{p}_{\tilde{\delta}}(\psi)$ for each $\psi \in USC(X)$, where $\tilde{p}_{\tilde{\delta}}(\psi)$ is defined in terms of the metric $\tilde{\rho}$. As the roles of ρ and $\tilde{\rho}$ in this argument can be interchanged, it follows that $\lim_{\delta \searrow 0} p_\psi(\delta)$ is a quantity that does not depend on the choice of a particular metric generating the topology on X. ◇

Proof of the theorem: The equality between the supremum and the limit is a consequence of the monotonicity of $\delta \mapsto p_\delta(\psi)$. As $p_\delta(\psi) \leq h(\mu) + \mu(\psi) \leq p(\psi)$

by Corollary 4.4.9, all we have to show is that $h(\mu) + \mu(\psi) \leq \sup_{\delta>0} p_\delta(\psi)$ for all $\mu \in \mathcal{M}(\mathcal{T})$. Because of the convex affinity of the entropy function under ergodic decompositions (Theorem 4.3.7) it suffices to consider ergodic μ.

So fix an ergodic $\mu \in \mathcal{M}(\mathcal{T})$. We have to show that for each $\epsilon > 0$ there is $\delta > 0$ such that

$$h(\mu) + \mu(\psi) \leq p_\delta(\psi) + \epsilon . \tag{4.24}$$

The proof, which is a generalization of the argument given for shift spaces in Example 4.4.5, requires a rather technical passage from finite measurable partitions to maximal (n, δ)-separated subsets of X. Our proof, although inspired by Misiurewicz's presentation in [41], is different from that in that it allows the function $\psi \in USC(X)$ to be discontinuous.

Let $\epsilon \in (0, 1)$ and let α be a finite partition of X. Denote the elements of the partition α by $A_1, \ldots, A_s$. Without loss of generality $s \geq 3$. Because of the regularity of μ (see A.4.28) there exists for each A_i a compact set $B_i \subseteq A_i$ such that

$$\mu(A_i \setminus B_i) \leq \frac{\epsilon^2}{2s \log s} .$$

Let $B_0 := X \setminus \bigcup_{i=1}^s B_i$. Then $\mu(B_0) \leq \frac{1}{2\log s}\epsilon^2 < \frac{1}{2}\epsilon^2$ and $\beta := \{B_0, \ldots, B_s\}$ is a finite partition of X. As $\mu(A_i \mid B_j) = \delta_{ij}$ for all $i, j = 1, \ldots, s$, we have

$$\begin{aligned}
H_\mu(\alpha \mid \beta) &= -\sum_{i=1}^{s} \int \mu(A_i \mid \beta) \log \mu(A_i \mid \beta) \, d\mu \\
&= -\sum_{i=1}^{s} \sum_{j=0}^{s} \mu(B_j) \mu(A_i \mid B_j) \log \mu(A_i \mid B_j) \\
&= -\mu(B_0) \cdot \sum_{i=1}^{s} \mu(A_i \mid B_0) \log \mu(A_i \mid B_0) \leq \mu(B_0) \cdot \log s \\
&= \log s \cdot \sum_{i=1}^{s} \mu(A_i \setminus B_i) \leq \frac{\epsilon^2}{2} .
\end{aligned}$$

Just as in the proof of Lemma 3.2.15 it follows that

$$H_\mu(\alpha^{\Lambda_n} \mid \beta^{\Lambda_n}) \leq \lambda_n \cdot H_\mu(\alpha \mid \beta) \leq \lambda_n \cdot \frac{\epsilon^2}{2} . \tag{4.25}$$

For each $B \in \beta^{\Lambda_n}$ there is a point $x_B \in B$ such that $\int_B \psi_n \, d\mu \leq \mu(B) \cdot \psi_n(x_B)$. Let $E := \{x_B : B \in \beta^{\Lambda_n}\}$. If the set E were (n, δ)-separated, we could continue as in the special case of shift spaces treated in Example 4.4.5. As this is not the case, we are going to delete from β^{Λ_n} a set of "bad" elements of small total μ-measure such that the family of points x_B belonging to the remaining "good" sets B can be split up into a relatively small number of (n, δ)-separated sets.

Let

$$X_n := \left\{ x \in X : \frac{1}{\lambda_n} \sum_{g \in \Lambda_n} 1_{B_0}(T^g x) > 2\epsilon^2 \right\} .$$

As

$$\lim_{n\to\infty} \frac{1}{\lambda_n} \sum_{g \in \Lambda_n} 1_{B_0} \circ T^g = \mu(B_0) \le \epsilon^2 \quad \mu\text{-a.s.}$$

by Birkhoff's ergodic theorem and the ergodicity of μ, we have

$$\lim_{n\to\infty} \mu(X_n) = 0 . \tag{4.26}$$

By its very definition, X_n is a union of elements of β^{Λ_n}. Let $\beta_0^{\Lambda_n} := \{B \in \beta^{\Lambda_n} : B \subset X_n\}$ and $\beta_1^{\Lambda_n} := \beta^{\Lambda_n} \setminus \beta_0^{\Lambda_n} = \{B \in \beta^{\Lambda_n} : B \cap X_n = \emptyset\}$. Then

$$\begin{aligned} H_\mu(\beta^{\Lambda_n}) + \lambda_n \cdot \mu(\psi) &= H_\mu(\beta^{\Lambda_n}) + \mu(\psi_n) \\ &\le \mu(X_n^c) \sum_{B \in \beta_1^{\Lambda_n}} \frac{\mu(B)}{\mu(X_n^c)} \log \frac{\exp \psi_n(x_B)}{\mu(B)} \\ &\quad - \mu(X_n) \sum_{B \in \beta_0^{\Lambda_n}} \frac{\mu(B)}{\mu(X_n)} \log \mu(B) + \mu(X_n) \sup \psi_n . \end{aligned}$$

Using Jensen's inequality as in (4.9) the first sum is estimated from above by

$$\mu(X_n^c) \log \sum_{B \in \beta_1^{\Lambda_n}} \frac{\exp \psi_n(x_B)}{\mu(X_n^c)} = \mu(X_n^c) \cdot \left(P(\psi_n, E \setminus X_n) - \log \mu(X_n^c) \right) .$$

The second sum is estimated from above by

$$\begin{aligned} & -\mu(X_n) \log \mu(X_n) - \mu(X_n) \sum_{B \in \beta_0^{\Lambda_n}} \frac{\mu(B)}{\mu(X_n)} \log \frac{\mu(B)}{\mu(X_n)} \\ \le\ & -\mu(X_n) \log \mu(X_n) + \mu(X_n) \log |\beta^{\Lambda_n}| . \end{aligned}$$

Hence

$$H_\mu(\beta^{\Lambda_n}) + \lambda_n \cdot \mu(\psi) \le P(\psi_n, E \setminus X_n) + \log 2 + \mu(X_n) \left(\sup \psi_n + \log |\beta^{\Lambda_n}| \right) .$$

We are going to show later that there is some $\delta > 0$ such that

$$P(\psi_n, E \setminus X_n) \le P_{n,\delta}(\psi) + \frac{\epsilon}{2} \lambda_n + O(\log \lambda_n) . \tag{4.27}$$

Assuming this inequality for the moment we can continue the above estimate:

$$\frac{1}{\lambda_n} H_\mu(\beta^{\Lambda_n}) + \mu(\psi) \le \frac{1}{\lambda_n} P_{n,\delta}(\psi) + \frac{\epsilon}{2} + O\left(\frac{\log \lambda_n}{\lambda_n} \right) + \mu(X_n)(\sup \psi + \log |\beta|) .$$

Combining this with (4.25) and observing Corollary 3.1.6-c and (4.26) this yields in the limit $n \to \infty$

$$\begin{aligned} h(\mu, \alpha, \mathcal{T}) + \mu(\psi) &= \lim_{n\to\infty} \frac{1}{\lambda_n} H_\mu(\alpha^{\Lambda_n}) + \mu(\psi) \\ &\leq \lim_{n\to\infty} \frac{1}{\lambda_n} H_\mu(\beta^{\Lambda_n}) + \mu(\psi) + \frac{\epsilon^2}{2} \\ &\leq p_\delta(\psi) + \epsilon\,. \end{aligned}$$

This is (4.24).

We still must show that there is some $\delta > 0$ satisfying (4.27). Choose $\delta := \frac{1}{2}\min\{\rho(B_i, B_j) : i, j \neq 0 \text{ and } i \neq j\}$. Let

$$\mathcal{W}_n := \{(\Gamma, w) : \Gamma \subset \Lambda_n, |\Gamma| \leq 2\epsilon^2\lambda_n, w \in \{1, \ldots, s\}^\Gamma\}\,.$$

As the number of subsets of Λ_n whose size is at most $2\epsilon^2\lambda_n$ does not exceed $2\epsilon^2\lambda_n\, e^{H(2\epsilon^2)\lambda_n + O(\log\lambda_n)}$ where $H(t)$ is again the entropy of the probability vector $(t, 1-t)$ (use estimate (1.7) from Example 1.2.1 based on Stirling's formula with $N = \lambda_n$, $\Sigma = \{0, 1\}$ and $k \in \{(\lambda_n - j, j) : 1 \leq j \leq 2\epsilon^2\lambda_n\}$), we have

$$|\mathcal{W}_n| \leq 2\epsilon^2\lambda_n\, e^{O(\log\lambda_n)}\, e^{H(2\epsilon^2)\lambda_n}\, s^{2\epsilon^2\lambda_n}$$

so that

$$\log|\mathcal{W}_n| \leq \lambda_n(H(2\epsilon^2) + 2\epsilon^2\log s) + O(\log\lambda_n)\,.$$

Each $(\Gamma, w) \in \mathcal{W}_n$ specifies a subset of $E \setminus X_n$:

$$E(\Gamma, w) := \{x \in E \setminus X_n : T^g x \in B_0 \text{ iff } g \in \Gamma;\ T^g x \in A_{w_g}\ \forall g \in \Gamma\}\,.$$

By definition of X_n we have $E \setminus X_n = \bigcup_{(\Gamma,w)\in\mathcal{W}_n} E(\Gamma, w)$. Fix some $(\Gamma, w) \in \mathcal{W}_n$ and consider $x_B, x_{B'} \in E(\Gamma, w)$ with $B \neq B'$. There must be some $g \in \Lambda_n$ such that $T^g x_B$ and $T^g x_{B'}$ belong to different elements of β. Obviously this $g \notin \Gamma$. Hence $T^g x_B$ and $T^g x_{B'}$ belong to two different sets B_i and B_j with $i, j \neq 0$. But then, by choice of δ, they have distance larger than δ. Therefore each set $E(\Gamma, w)$ is (n, δ)-separated, and it follows that

$$\exp P(\psi_n, E \setminus X_n) \leq \sum_{(\Gamma,w)\in\mathcal{W}_n} \sum_{x\in E(\Gamma,w)} e^{\psi_n(x)} \leq |\mathcal{W}_n| \cdot \exp P_{n,\delta}(\psi)\,.$$

Hence

$$\begin{aligned} P(\psi_n, E \setminus X_n) &\leq \log|\mathcal{W}_n| + P_{n,\delta}(\psi) \\ &\leq P_{n,\delta}(\psi) + \lambda_n(H(2\epsilon^2) + 2\epsilon^2\log s) + O(\log\lambda_n)\,. \end{aligned}$$

As $H(2\epsilon^2) + 2\epsilon^2\log s < \epsilon$ if $\epsilon > 0$ is small enough, (4.24) follows for $n \to \infty$. □

The previous results have the following immediate implication for the set of constructible δ-equilibrium states $CES(\delta, \psi, \mathcal{T})$:

4.4.13 Corollary *Let $\mathcal{T}$ be a continuous action of $G = \mathbb{Z}^d$ or $G = \mathbb{Z}_+^d$ on the compact metric space (X, ρ) and suppose that $\psi \in USC(X)$. If $p_{\delta_0}(\psi) = \sup_{\delta>0} p_\delta(\psi)$ for some $\delta_0 > 0$, then $CES(\delta_0, \psi, \mathcal{T}) \subseteq ES(\psi, \mathcal{T})$.*

Proof: By Corollary 4.4.9 and Theorem 4.4.11 we have for $\mu \in CES(\delta_0, \psi, \mathcal{T})$

$$h(\mu) + \mu(\psi) \geq p_{\delta_0}(\psi) = p(\psi) .$$

□

4.5 Equilibrium states for expansive actions

The last corollary raises the questions which systems do have the property that $p_{\delta_0}(\psi) = \sup_{\delta>0} p_\delta(\psi)$ for some $\delta_0 > 0$ and to what extent $CES(\delta_0, \psi, \mathcal{T})$ does exhaust $ES(\psi, \mathcal{T})$ in this case. We are going to show that *expansive* systems (defined below) have the property in question and that for such systems $ES(\psi, \mathcal{T})$ is the closed convex hull of $CES(\delta, \psi, \mathcal{T})$.

4.5.1 Definition *The action $\mathcal{T}$ on (X, ρ) is expansive with expansiveness constant $\delta > 0$, if for any $x, y \in X$ with $x \neq y$ there exists an element $g \in G$ such that $\rho(T^g x, T^g y) \geq \delta$.*

4.5.2 Remark If $\mathcal{T}$ is expansive on (X, ρ) and if $\tilde{\rho}$ is an equivalent metric on X, then $\mathcal{T}$ is expansive on $(X, \tilde{\rho})$ with some expansiveness constant $\tilde{\delta}$, see Remark 4.4.12. Therefore the property of an action $\mathcal{T}$ on a compact metrizable space X of being expansive does not depend on the choice of a particular metric for X. ◇

4.5.3 Example The shift action $\mathcal{T}$ on $\Omega = \Sigma^G$ is expansive with expansiveness constant 1. For if $\omega \neq \omega'$, then there exists $g \in G$ such that $(T^g\omega)_0 = \omega_g \neq \omega'_g = (T^g\omega')_0$ and hence $d(T^g\omega, T^g\omega') = 1$. (The metric d was defined in Example 4.2.2.) ◇

4.5.4 Lemma *If $\mathcal{T}$ is expansive with expansiveness constant $\delta_0 > 0$, then for every $\delta \in (0, \delta_0)$ there exists an index $k(\delta) \in \mathbb{N}$ such that $U(k(\delta), \delta_0, x) \subseteq B_\delta(x)$ for all $x \in X$.*

Proof: Suppose for a contradiction that there is $\delta \in (0, \delta_0)$ for which there are points $x_k, y_k \in X$ $(k \in \mathbb{N})$ such that $\rho(x_k, y_k) \geq \delta$ but $\rho(T^g x_k, T^g y_k) < \delta_0$ for all $g \in \Lambda_k$. As X is compact there are subsequences (x_{k_i}) and (y_{k_i}) converging to points x and y in X respectively, so that $\rho(x, y) \geq \delta > 0$ and $\rho(T^g x, T^g y) \leq \delta_0$ for all $g \in \bigcup_i \Lambda_{k_i} = G$. This contradicts the expansiveness of $\mathcal{T}$. □

4.5.5 Corollary *Suppose that $\mathcal{T}$ is an expansive action with expansiveness constant $\delta_0 > 0$.*

1. *If $\delta \in (0, \delta_0)$, then each (n, δ)-separated set is $(n + k(\delta), \delta_0)$-separated.*
2. *$p_{\delta_0}(\psi) = \sup_{\delta>0} p_\delta(\psi)$ for each $\psi \in USC_0(X)$ (i.e., such that $\{x : \psi(x) = -\infty\}$ is $\mathcal{T}$-invariant and $\underline{\psi} := \inf\{\psi(x) : \psi(x) > -\infty\} > -\infty$, cf. Definition 4.4.4).*

Proof: The first assertion follows from Lemma 4.5.4, because

$$\begin{aligned} U(n + k(\delta), \delta_0, x) &= \bigcap_{g\in\Lambda_n} T^{-g}U(k(\delta), \delta_0, T^g x) \\ &\subseteq \bigcap_{g\in\Lambda_n} T^{-g}B_\delta(T^g x) = U(n, \delta, x) \end{aligned}$$

for all x. It implies in particular that $P_{n,\delta}(\psi) \le P_{n+k(\delta),\delta_0}(\psi) + (\lambda_{n+k(\delta)} - \lambda_n)|\underline{\psi}|$ for each $\delta \in (0, \delta_0)$. Hence

$$\begin{aligned} p_\delta(\psi) &= \limsup_{n\to\infty} \frac{1}{\lambda_n} P_{n,\delta}(\psi) \\ &\le \limsup_{n\to\infty} \frac{\lambda_{n+k(\delta)}}{\lambda_n} \cdot \frac{1}{\lambda_{n+k(\delta)}} P_{n+k(\delta),\delta_0}(\psi) + \limsup_{n\to\infty} \frac{\lambda_{n+k(\delta)} - \lambda_n}{\lambda_n} |\underline{\psi}| \\ &= p_{\delta_0}(\psi)\,, \end{aligned}$$

and the second assertion follows from the monotonicity of $\delta \mapsto p_\delta(\psi)$. □

Expansive actions fit nicely into the general framework described in the last two sections where the upper semicontinuity of the entropy function played an important role:

4.5.6 Theorem (Semicontinuity of the entropy for expansive actions)
Suppose $\mathcal{T}$ is an expansive action on X. Then the entropy function h is upper semicontinuous.

Proof: Let $\delta_0 > 0$ be an expansiveness constant for $\mathcal{T}$ and suppose that $\mu_n \in \mathcal{M}(\mathcal{T})$ is a sequence converging to μ. By Lemma 4.4.10 there exists a finite partition α such that diam$(A) < \delta_0$ and $\mu(\partial A) = 0$ for all $A \in \alpha$. In view of Theorem 4.2.4 it suffices to show that α is a generator for $\mathcal{T}$. Denote by $A_k(x)$ the unique element from α^{Λ_k} that contains x. Then $A_k(x) \subseteq U(k, \delta_0, x)$ for all k and all $x \in X$. Given $\delta > 0$ choose $k(\delta)$ as in Lemma 4.5.4. Then

$$A_{k(\delta)}(x) \subseteq U(k(\delta), \delta_0, x) \subseteq B_\delta(x)\,. \tag{4.28}$$

Denote $\mathcal{A} = \bigcup_{k=0}^\infty \alpha^{\Lambda_k}$ and let $V \subset X$ be any open set. Because of (4.28) we have $V = \bigcup_{A\in\mathcal{A}, A\subseteq V} A$, and as $\mathcal{A}$ is at most countable, it follows that $V \in \sigma(\mathcal{A})$. Hence

$\mathcal{B} = \sigma(\mathcal{A})$, i.e., α is a generator. □

The following lemma not only is the essential technical step in the proof of the fact that $ES(\psi, \mathcal{T})$ is the closed convex hull of $CES(\delta_0, \psi, \mathcal{T})$ (see the next theorem), but will also be useful for the investigation of the Ising model in Chapter 5.

4.5.7 Lemma *Let $\mathcal{T}$ be a continuous action of $G = \mathbb{Z}^d$ or $G = \mathbb{Z}_+^d$ on the compact metric space (X, ρ) and suppose that $\psi \in USC(X)$, $\mu \in ES(\psi, \mathcal{T})$ and $\phi \in C(X)$. If there is some $\delta_0 > 0$ such that $p_{\delta_0}(\tilde{\psi}) = \sup_{\delta>0} p_\delta(\tilde{\psi})$ for all $\tilde{\psi} = \psi + \tilde{\phi}$ with $\tilde{\phi} \in C(X)$, then there exists $\mu_\phi \in CES(\delta_0, \psi, \mathcal{T})$ such that $\mu(\phi) \le \mu_\phi(\phi)$.*

Proof: By Remark 4.3.2-b, $ES(\psi, \mathcal{T}) \subseteq \mathcal{D}_\psi(p)$ so that

$$p_{\delta_0}(\psi + t\phi) = p(\psi + t\phi) \ge p(\psi) + t\mu(\phi) = p_{\delta_0}(\psi) + t\mu(\phi)$$

for all $t \in \mathbb{R}$. Let $0 < t_k \to 0$ as $k \to \infty$. Then for each $k \in \mathbb{N}$ there exists $n(k) \in \mathbb{N}$ such that

$$p(\psi + t_k\phi) \quad \le \quad \frac{1}{\lambda_{n(k)}} P_{n(k),\delta_0}(\psi + t_k\phi) + t_k^2$$

and

$$p(\psi) \quad \ge \quad \frac{1}{\lambda_{n(k)}} P_{n(k),\delta_0}(\psi) - t_k^2 \,.$$

(For later use we note that we may choose $n(k) \ge t_k^{-2}$.) For each $n(k)$ there exists an $(n(k), \delta_0)$-separated set $E_{n(k)}$ such that

$$P_{n(k),\delta_0}(\psi + t_k\phi) \le P(\psi_{n(k)} + t_k\phi_{n(k)}, E_{n(k)}) + t_k^2\lambda_{n(k)} \tag{4.29}$$

and by definition

$$P_{n(k),\delta_0}(\psi) \ge P(\psi_{n(k)}, E_{n(k)}) \,.$$

Hence

$$\begin{aligned} t_k\mu(\phi) &\le p(\psi + t_k\phi) - p(\psi) \\ &\le \frac{1}{\lambda_{n(k)}} \left(P_{n(k),\delta_0}(\psi + t_k\phi) - P_{n(k),\delta_0}(\psi)\right) + 2t_k^2 \\ &\le \frac{1}{\lambda_{n(k)}} \left(P(\psi_{n(k)} + t_k\phi_{n(k)}, E_{n(k)}) - P(\psi_{n(k)}, E_{n(k)})\right) + 3t_k^2 \,. \end{aligned}$$

Let the measures $\pi_n = \sum_{z\in E_n} \delta_z \cdot \exp(\psi_n(z) - P(\psi_n, E_n))$ be as in (4.12). Then

$$\begin{aligned} &P(\psi_{n(k)} + t_k\phi_{n(k)}, E_{n(k)}) - P(\psi_{n(k)}, E_{n(k)}) \\ &= \log \pi_{n(k)}(e^{t_k\phi_{n(k)}}) \\ &= t_k \cdot \frac{\pi_{n(k)}(\phi_{n(k)}\, e^{\tilde{t}_k\phi_{n(k)}})}{\pi_{n(k)}(e^{\tilde{t}_k\phi_{n(k)}})} \end{aligned}$$

for some $\tilde{t}_k \in (0, t_k)$ by the mean value theorem of differentiation. Combining the last two estimates we arrive at

$$\mu(\phi) \leq \frac{1}{\lambda_{n(k)}} \frac{\pi_{n(k)}(\phi_{n(k)}\, e^{\tilde{t}_k \phi_{n(k)}})}{\pi_{n(k)}(e^{\tilde{t}_k \phi_{n(k)}})} + 3t_k . \tag{4.30}$$

Let

$$\psi^{(n(k))} := \psi_{n(k)} + \tilde{t}_k \phi_{n(k)} .$$

Then

$$\|\psi^{(n(k))} - \psi_{n(k)}\|_\infty \leq t_k \lambda_{n(k)} \|\phi\|_\infty = o(\lambda_{n(k)})$$

so that the functions $\psi^{(n(k))}$ satisfy (4.10) and define measures

$$\tilde{\pi}_{n(k)} = \sum_{z \in E_{n(k)}} \delta_z \cdot \exp\left(\psi^{(n(k))}(z) - P(\psi^{(n(k))}, E_{n(k)})\right)$$

as in (4.12). Then

$$\frac{\pi_{n(k)}(\phi_{n(k)}\, e^{\tilde{t}_k \phi_{n(k)}})}{\pi_{n(k)}(e^{\tilde{t}_k \phi_{n(k)}})} = \tilde{\pi}_{n(k)}(\phi_{n(k)}) ,$$

and observing the relation (4.13) on p. 81 between the measures $\tilde{\pi}_{n(k)}$ and the corresponding measures $\tilde{\mu}_{n(k)}$ we conclude from (4.30) that

$$\mu(\phi) \leq \frac{1}{\lambda_{n(k)}} \tilde{\pi}_{n(k)}(\phi_{n(k)}) + 3t_k = \tilde{\mu}_{n(k)}(\phi) + 3t_k .$$

Passing to a subsequence, if necessary, the sequence $(\tilde{\mu}_{n(k)})_k$ converges to some $\mu_\phi \in \mathcal{M}(\mathcal{T})$ with

$$\mu(\phi) \leq \mu_\phi(\phi) .$$

In order to guarantee that $\mu_\phi \in CES(\delta_0, \psi, \mathcal{T})$ we must check (4.14) and (4.15) on p. 82. But in view of the choice of the $n(k)$ and of the sets $E_{n(k)}$, elementary estimates lead to

$$\begin{aligned} P(\psi_{n(k)}, E_{n(k)}) &\geq P(\psi_{n(k)} + t_k \phi_{n(k)}, E_{n(k)}) - t_k \lambda_{n(k)} \|\phi\|_\infty \\ &\geq P_{n(k),\delta_0}(\psi + t_k \phi) - o(\lambda_{n(k)}) . \end{aligned}$$

As

$$P_{n(k),\delta_0}(\psi + t_k \phi) \geq P_{n(k),\delta_0}(\psi) - t_k \lambda_{n(k)} \|\phi\|_\infty$$

this gives (4.15), and (4.14) follows from

$$P_{n(k),\delta_0}(\psi + t_k \phi) \geq \lambda_{n(k)} p(\psi + t_k \phi) - o(\lambda_{n(k)}) \geq \lambda_{n(k)} p(\psi) - o(\lambda_{n(k)}) .$$

□

4.5.8 Remark In Chapter 5 we shall refer to the preceding lemma in the more specific setting $X = \Omega = \Sigma^{\mathbb{Z}^d}$ and $\delta_0 = 1$. For the local energy functions ψ and ϕ there will be a constant $C_{\psi,\phi} > 0$ such that

$$|P(\psi_n + t\phi_n, E_n) - P(\psi_n + t\phi_n, E'_n)| \leq C_{\psi,\phi}\, n^{d-1} \tag{4.31}$$

for any $n > 0$, $|t| \leq 1$ and any maximal $(n, 1)$-separated sets $E_n, E'_n \subset \Omega$. This allows us to use only very specific sets $E_{n(k)}$ in inequality (4.29) of the preceding proof: We fix some $\xi \in \Omega$ and let $E_n^\xi = \{\omega \in \Omega : \omega_g = \xi_g \ \forall g \in G \setminus \Lambda_n\}$. As $n(k) \geq t_k^{-2}$ by choice of the $n(k)$, we have

$$\begin{aligned} &|P(\psi_{n(k)} + t_k\phi_{n(k)}, E_{n(k)}) - P(\psi_{n(k)} + t_k\phi_{n(k)}, E^\xi_{n(k)})| \\ &\leq \ C_{\psi,\phi}\, n(k)^{d-1} \leq C_{\psi,\phi} t_k^2 \lambda_{n(k)} \,. \end{aligned}$$

Therefore the sets $E_{n(k)}$ in (4.29) can always be chosen as $E^\xi_{n(k)}$. This yields a measure $\mu^\xi_\phi \in \mathit{CES}(1, \psi, \mathcal{T})$ satisfying $\mu(\phi) \leq \mu^\xi_\phi(\phi)$. (The measure μ^ξ_ϕ, though, need not coincide with a μ_ϕ obtained from a less restrictive choice of the $E_{n(k)}$.)

We express the fact that μ^ξ_ϕ was constructed using only the special sets $E^\xi_{n(k)}$ and weight functions $\psi^{(n(k))} = \psi_{n(k)} + \tilde{t}_k\phi_{n(k)}$ with $\tilde{t}_k > 0$ and ϕ and ψ satisfying (4.31) by saying $\mu^\xi_\phi \in \mathit{CES}^\xi_\phi(1, \psi, \mathcal{T})$. ◇

4.5.9 Theorem (Equilibrium states by explicit construction)
Let $\mathcal{T}$ be a continuous action of $G = \mathbb{Z}^d$ or $G = \mathbb{Z}^d_+$ on the compact metric space (X, ρ) and suppose that $\psi \in USC(X)$. If there is some $\delta_0 > 0$ such that $p_{\delta_0}(\tilde{\psi}) = \sup_{\delta > 0} p_\delta(\tilde{\psi})$ for all $\tilde{\psi} = \psi + \tilde{\phi}$ with $\tilde{\phi} \in C(X)$, then $ES(\psi, \mathcal{T})$ is the closed convex hull of $CES(\delta_0, \psi, \mathcal{T})$. (The assumption involving δ_0 is satisfied if $\mathcal{T}$ is expansive and $\psi \in USC_0(X)$, see Corollary 4.5.5.)

Proof: Denote by K the closed convex hull of $\mathit{CES}(\delta_0, \psi, \mathcal{T})$ in $\mathcal{M}(\mathcal{T})$ (with respect to the weak topology). We know that $K \subseteq \mathit{ES}(\psi, \mathcal{T})$ by Corollary 4.4.13 and Theorem 4.2.3. Suppose that there exists some $\mu \in \mathit{ES}(\psi, \mathcal{T}) \setminus K$. Then μ has an open neighbourhood that is disjoint from K. In particular, there are $\phi \in C(X)$ and $\epsilon > 0$ such that

$$\mu(\phi) > \nu(\phi) + \epsilon \quad \text{for all } \nu \in K, \tag{4.32}$$

compare the proof of Theorem 4.2.9. On the other hand, because of Lemma 4.5.7 there is a measure $\mu_\phi \in \mathit{CES}(\delta_0, \psi, \mathcal{T}) \subseteq K$ such that $\mu(\phi) \leq \mu_\phi(\phi)$. This contradicts (4.32). □

4.5.10 Remark In the special setting of Remark 4.5.8 assume additionally that there is a $\|\,.\,\|_\infty$-dense set $D \subset C(X)$ of functions ϕ that satisfy (4.31). Denote by K^ξ the closed convex hull of $\bigcup_{\phi \in D} \mathit{CES}^\xi_\phi(1, \psi, \mathcal{T})$, and suppose that there is some $\mu \in \mathit{ES}(\psi, \mathcal{T}) \setminus K^\xi$. Then there is some $\phi \in D$ that separates μ from K^ξ and exactly the same arguments as in the proof of the preceding theorem lead to the contradiction $\mu^\xi_\phi(\phi) + \epsilon < \mu(\phi) \leq \mu^\xi_\phi(\phi)$ for a measure $\mu^\xi_\phi \in CES^\xi_\phi(1, \psi, \mathcal{T})$. Therefore $ES(\psi, \mathcal{T})$ is the closed convex hull of $\bigcup_{\phi \in D} CES^\xi_\phi(1, \psi, \mathcal{T})$. ◇

Chapter 5

Gibbs measures

In Section 1.1 we were able to identify equilibrium states on finite configuration spaces Ω as *Gibbs measures* (or *Gibbs states*). That is, we showed that for a given energy function $u : \Omega \to \mathbb{R}$ the unique measure μ that maximizes $H(\mu) + \mu(u)$ has the form $\mu(\omega) = \exp(u(\omega) - p(u))$ where $p(u) = \log \sum_{\omega \in \Omega} e^{u(\omega)}$. (In fact, in the notation of Section 1.1, $\mu = \mu_\beta$ for $\beta = -1$.) In Section 1.2 we specialized this to configuration spaces $\Omega = \Sigma^G$ where G is a finite lattice and $u = \sum_{g \in G} \psi \circ T^g$. Now the measure μ maximizing $\frac{1}{|G|} H(\mu) + \mu(\psi)$ has the form

$$\mu(\omega) = \exp\left(\sum_{g \in G} (\psi(T^g \omega) - p(\psi))\right) \tag{5.1}$$

where $p(\psi) = \frac{1}{|G|} \log \sum_{\omega \in \Omega} e^{u(\omega)}$.

Evidently such an explicit representation cannot be expected for Gibbs states on infinite lattices, because there each single configuration has measure 0. However, if we fix a configuration outside a coordinate box Λ_n, we can ask for the conditional distribution of an equilibrium state inside this finite box given the values of the configuration outside. The goal of this chapter is to show that for sufficiently regular local energy functions ψ these conditional distributions can be described in a way that bears some resemblance to (5.1).

In this whole chapter we restrict ourselves to the case $X = \Omega = \Sigma^G$ where $G = \mathbb{Z}^d$, for which one of the classical sources is [52]. Readers interested in Gibbs theory for more general actions are referred to the literature, for example to [28], [53].

5.1 Regular local energy functions

In this section we describe a class of local energy functions ψ for which the effect on the values $\psi_n(\omega)$ of changing a configuration ω at a single site can be controlled uniformly in n.

5.1.1 Definition *Let $\psi : \Omega \to \mathbb{R}$ be a local energy function.*

1. *ψ is regular, if*

$$\sum_{n=1}^{\infty} n^{d-1} \cdot \delta_n(\psi) < \infty \tag{5.2}$$

where $\delta_n(\psi) := \sup\{|\psi(\omega) - \psi(\omega')| : \omega, \omega' \in \Omega,\ \omega_g = \omega'_g\ \forall g \in \Lambda_n\}$.

2. *ψ has finite range, if there is some $n_0 \in \mathbb{N}$ such that $\delta_n(\psi) = 0$ for all $n \geq n_0$.*

5.1.2 Remarks

a) If ψ has finite range, then it is regular, and if ψ is regular, then it is continuous. (In fact, $\lim_{n\to\infty} \delta_n(\psi) = 0$ is equivalent to the continuity of ψ.)

b) The set of local energy functions of finite range (and a fortiori that of regular ones) is $\|\,.\,\|_\infty$-dense in $C(\Omega)$.

c) If $\psi : \Omega \to \mathbb{R}$ is Hölder-continuous, then it is regular. ◇

5.1.3 Example Let $\Sigma = \{-1, 1\}$ and denote by $E \subset G$ the set of canonical unit vectors in G. Then the local energy function $\psi(\omega) := \sum_{e\in E}(\omega_0\omega_e + \omega_0\omega_{-e})$ which defines the Ising model (see Example 1.2.2) is regular. In fact, ψ has finite range since $\delta_n(\psi) = 0$ for $n \geq 2$. ◇

5.1.4 Lemma *Consider $\psi : \Omega \to \mathbb{R}$ and denote by α the canonical partition of Ω. Let $\Delta_n(\psi) := \sup\{|\psi_n(\omega) - \psi_n(\omega')| : \omega, \omega' \in A,\ A \in \alpha^{\Lambda_n}\}$.*

1. *If $\psi \in C(\Omega)$, then $\lim_{n\to\infty} \frac{1}{\lambda_n}\Delta_n(\psi) = 0$.*
2. *If ψ is regular and $C := \sum_{j=1}^{\infty} \delta_j(\psi)$, then*

$$\Delta_n(\psi) \leq 2dC\,(2n-1)^{d-1}\,. \tag{5.3}$$

If E_n and E'_n are two maximal $(n,1)$-separated subsets of Ω, then

$$|P(\psi_n, E_n) - P(\psi_n, E'_n)| \leq 2dC\,(2n-1)^{d-1}\,. \tag{5.4}$$

Proof: Let $\omega, \omega' \in A \in \alpha^{\Lambda_n}$. Then

$$\begin{aligned} |\psi_n(\omega) - \psi_n(\omega')| &\leq \sum_{g\in\Lambda_n} |\psi(T^g\omega) - \psi(T^g\omega')| \\ &\leq \sum_{j=0}^{n-1} \sum_{g\in\Lambda_{j+1}\setminus\Lambda_j} \delta_{n-j}(\psi) \\ &\leq 2d \sum_{j=0}^{n-1} (2j+1)^{d-1}\delta_{n-j}(\psi) \end{aligned} \tag{5.5}$$

as $|\Lambda_{j+1} \setminus \Lambda_j| \leq 2d\,(2j+1)^{d-1}$. If ψ is regular, (5.3) follows immediately.

As $|A \cap E_n| = |A \cap E'_n| = 1$ for each $A \in \alpha^{\Lambda_n}$ (see Example 4.4.2) there are, for each $A \in \alpha^{\Lambda_n}$, configurations $\omega_A, \omega'_A \in A$ such that $E_n = \{\omega_A : A \in \alpha^{\Lambda_n}\}$ and $E'_n = \{\omega'_A : A \in \alpha^{\Lambda_n}\}$. Now (5.4) follows from (5.3), because

$$\begin{aligned} P(\psi_n, E_n) &= \log \sum_{A \in \alpha^{\Lambda_n}} \exp \psi_n(\omega_A) \\ &\le \log \left(\sum_{A \in \alpha^{\Lambda_n}} \exp \psi_n(\omega'_A) \exp(2dC\,(2n-1)^{d-1}) \right) \\ &= P(\psi_n, E'_n) + 2dC\,(2n-1)^{d-1} \end{aligned}$$

and vice versa.

The first assertion of the lemma still remains to be proved. So we continue estimate (5.5) for arbitrary continuous ψ:

$$\begin{aligned} \Delta_n(\psi) &\le 2d\,2^{d-1} \sum_{j=0}^{n-[\sqrt{n}]} (j+1)^{d-1}\,\delta_{[\sqrt{n}]}(\psi) + 2d\,2^{d-1}\,[\sqrt{n}]\,n^{d-1}\,\delta_1(\psi) \\ &\le \text{const} \cdot (n^d\,\delta_{[\sqrt{n}]}(\psi) + n^{d-\frac{1}{2}}\,\delta_1(\psi)) = o(\lambda_n)\,. \end{aligned}$$

□

In order to study how $\psi_n(\omega)$ changes if ω is altered at finitely many sites, we introduce the following class of "local" homeomorphisms on Ω. Denote by $\mathcal{E}_n$ the set of all maps $\tau : \Omega \to \Omega$ such that

$$(\tau(\omega))_i = \begin{cases} \tau_i(\omega_i) & (i \in \Lambda_n), \\ \omega_i & (i \notin \Lambda_n), \end{cases}$$

where the maps $\tau_i : \Sigma \to \Sigma$ are permutations ($i \in \Lambda_n$), and let $\mathcal{E} = \bigcup_{n>0} \mathcal{E}_n$. Obviously $\mathcal{E}$ is a group of homeomorphisms of Ω that affect only finitely many coordinates.

5.1.5 Remarks

a) Observe that for $g \in G$ the following holds: $|g| = j$ if and only if $g \in \Lambda_{j+1} \setminus \Lambda_j$.

b) Recall that the group $G = \mathbb{Z}^d$ acts on Ω canonically by $(T^g\omega)_i = \omega_{i+g}$. One should note that T^g shifts the configuration ω by the vector $-g$ on the lattice G. In particular, $T^g([\sigma_i]_{i \in \Lambda_n}) = [\sigma_{i+g}]_{i \in \Lambda_n - g}$. Hence, if $\tau \in \mathcal{E}_n$, $g \in G$ and $i \in G \setminus (\Lambda_n - g)$, then

$$(T^g(\tau\omega))_i = (\tau\omega)_{i+g} = \omega_{i+g}\,.$$

This holds in particular if $|i| \le |g| - n$ or, equivalently, if $i \in \Lambda_{|g|+1-n}$. ◇

5.1.6 Lemma *Suppose $\psi : \Omega \to \mathbb{R}$ is regular. For $\tau \in \mathcal{E}$ and $n > 0$ define*

$$\Psi_\tau^n : \Omega \to \mathbb{R}, \quad \Psi_\tau^n := \psi_n \circ \tau^{-1} - \psi_n\,,$$

where as usual $\psi_n = \sum_{g\in\Lambda_n} \psi \circ T^g$. Then the limit

$$\Psi_\tau := \lim_{n\to\infty} \Psi_\tau^n \quad \text{exists uniformly on } \Omega$$

(in particular Ψ_τ is continuous), and for each $n_0 > 0$ the convergence is even uniform for all $\tau \in \mathcal{E}_{n_0}$. Furthermore,

$$\sup\{\|\Psi_\tau - \Psi_\tau^n\|_\infty \cdot n^{-(d-1)} : n > 0, \tau \in \mathcal{E}_n\} < \infty\,. \tag{5.6}$$

This lemma tells us that the asymptotic behaviour (as $n \to \infty$) of the "global" energy $\psi_n(\omega)$ is essentially independent of the value of ω_i at a single position $i \in G$.

Proof of the lemma: Let $0 < n_0 \le j$ and $\tau \in \mathcal{E}_{n_0}$. For $g \in \Lambda_{j+1} \setminus \Lambda_j$ we have because of Remark 5.1.5-b

$$(T^g(\tau\omega))_i = \omega_{i+g} = (T^g\omega)_i \quad \forall\omega \in \Omega\ \forall i \in \Lambda_{j+1-n_0}$$

so that $|\psi(T^g(\tau\omega)) - \psi(T^g\omega)| \le \delta_{j+1-n_0}(\psi)$ for all $\omega \in \Omega$. For $0 < n_0 \le n < m$ this implies

$$\begin{aligned}
\|\Psi_\tau^m - \Psi_\tau^n\|_\infty &\le \sum_{g\in\Lambda_m\setminus\Lambda_n} \|\psi \circ T^g \circ \tau^{-1} - \psi \circ T^g\|_\infty \\
&\le \sum_{j=n}^{m-1} \sum_{g\in\Lambda_{j+1}\setminus\Lambda_j} \delta_{j+1-n_0}(\psi) \\
&\le \sum_{j=n}^{m-1} \underbrace{|\Lambda_{j+1} \setminus \Lambda_j|}_{\le 2d(2j+1)^{d-1}} \delta_{j+1-n_0}(\psi) \\
&\le \sum_{j=n-n_0}^{\infty} 2d(2j + 2n_0 + 1)^{d-1}\delta_{j+1}(\psi) \\
&\to 0 \quad \text{as } n \to \infty
\end{aligned}$$

because of (5.2) and as $(2j + 2n_0 + 1) \le 4j$ if $j > n_0$. Observe that the value of the sum does not depend on $\tau \in \mathcal{E}_{n_0}$ for fixed n_0. This proves the first part of the lemma.

In order to derive (5.6) from the previous estimate observe that for $m \ge n = n_0$

$$\begin{aligned}
\|\Psi_\tau^m - \Psi_\tau^n\|_\infty &\le 2^d d \sum_{j=0}^{\infty} (j + n + 1)^{d-1}\delta_{j+1}(\psi) \\
&\le 2^d d \left((2n+1)^{d-1} \sum_{j=0}^{n} \delta_{j+1}(\psi) + \sum_{j=n+1}^{\infty} (2j)^{d-1}\delta_{j+1}(\psi) \right) \\
&= O(n^{d-1})
\end{aligned}$$

uniformly in m so that $\|\Psi_\tau - \Psi_\tau^n\|_\infty = O(n^{d-1})$ also. □

5.2 Gibbs measures are equilibrium states

In this section we introduce Gibbs measures for regular local energy functions and show that they are equilibrium states.

5.2.1 Definition *Let ψ be a regular local energy function. A Borel probability measure μ on Ω is a Gibbs measure (or Gibbs state) for ψ, if*

$$\tau\mu = \mu \cdot e^{\Psi_\tau}$$

for each $\tau \in \mathcal{E}$. The set of Gibbs states for ψ is denoted by $GS(\psi)$, the set of $\mathcal{T}$-invariant Gibbs states by $GS(\psi, \mathcal{T})$.

Although the proof that Gibbs measures really exist is postponed to the next section where we finish the proof that $GS(\psi, \mathcal{T}) = ES(\psi, \mathcal{T}) \neq \emptyset$, we collect a number of properties of Gibbs measures in this section.

5.2.2 Lemma *$GS(\psi)$ and $GS(\psi, \mathcal{T})$ are closed convex subsets of the set of all Borel probability measures on Ω.*

Proof: Let μ_n be a sequence in $GS(\psi)$ converging to some measure μ and let $\tau \in \mathcal{E}$. As τ and Ψ_τ are continuous, it follows that

$$\tau\mu = \lim_{n\to\infty} \tau\mu_n = \lim_{n\to\infty} \mu_n \cdot e^{\Psi_\tau} = \mu \cdot e^{\Psi_\tau} .$$

Hence $GS(\psi)$ is closed. It is convex as for $\mu_1, \mu_2 \in GS(\psi)$ and $a, b \in (0, 1)$ with $a + b = 1$ the following holds:

$$\tau(a\mu_1 + b\mu_2) = a(\tau\mu_1) + b(\tau\mu_2) = a(\mu_1 \cdot e^{\Psi_\tau}) + b(\mu_2 \cdot e^{\Psi_\tau}) = (a\mu_1 + b\mu_2) \cdot e^{\Psi_\tau} .$$

This proves the lemma for $GS(\psi)$. As $GS(\psi, \mathcal{T}) = GS(\psi) \cap \mathcal{M}(\mathcal{T})$, the claim for $GS(\psi, \mathcal{T})$ follows. □

The definition of Gibbs states given above goes back to Capocaccia [12] and does not involve conditional distributions. In order to link it to the more classical one, we must introduce some more notation.

1. Configurations $\omega \in \Omega$ are written as $\omega = (z, \eta)$ where $z \in \Omega_n := \Sigma^{\Lambda_n}$ and $\eta \in \tilde{\Omega}_n := \Sigma^{G \setminus \Lambda_n}$. The index n to which the decomposition $\omega = (z, \eta)$ refers will be clear from the context.
2. We fix some reference symbol $\sigma_0 \in \Sigma$ and denote $\underline{z} := \{\sigma_0\}^{\Lambda_n}$.

3. For $z \in \Omega_n$ we use the shorthand notation $[z] := \{\omega \in \Omega : \omega_g = z_g \, \forall g \in \Lambda_n\}$ for the cylinder set $[z_g]_{g \in \Lambda_n}$. As before, α is the canonical partition of Ω into cylinder sets $[\sigma]_0$, $\sigma \in \Sigma$. Let $A_n(\omega)$ denote the unique element of α^{Λ_n} that contains ω. Then $A_n(z, \eta) = [z]$.
4. For each $z \in \Omega_n$ we fix a map $\tau_z \in \mathcal{E}_n$ such that $\tau_z(z, \eta) = (\underline{z}, \eta)$. ($\tau_z$ is not uniquely determined in this way.)
5. By $\mathcal{H}_n \subset \mathcal{B}$ we denote the σ-algebra generated by the coordinate mappings $\omega \mapsto \omega_g$ ($g \in G \setminus \Lambda_n$). Keeping in mind that $\mu([z] \mid \mathcal{H}_n)(\omega)$ depends on $\omega = (z', \eta)$ only via η we are going to show in the next theorem that $Q_n : \Omega_n \times \bar{\Omega}_n \to [0, 1]$,

$$Q_n(z, \eta) := \frac{\exp \Psi_{\tau_z}(\underline{z}, \eta)}{\sum_{z' \in \Omega_n} \exp \Psi_{\tau_{z'}}(\underline{z}, \eta)},$$

is a version of $\mu([z] \mid \mathcal{H}_n)(\omega)$.

5.2.3 Remark Observe that for fixed η the function $z \mapsto Q_n(z, \eta)$ is an elementary Gibbs measure in the sense of Section 1.1. In contrast to the measures π_n used in the constructive approach to equilibrium states in (4.12) we are now dealing with *conditional* Gibbs measures. ◇

5.2.4 Theorem (Characterization of Gibbs measures)
Let ψ be a regular local energy function.

a) *For all $\mu \in GS(\psi)$, $z \in \Omega_n$ and μ-a.e. $\omega = (z', \eta) \in \Omega$*

$$\mu([z] \mid \mathcal{H}_n)(\omega) = Q_n(z, \eta), \tag{5.7}$$

i.e., Q_n (and hence ψ) determines uniquely the conditional distributions of all $\mu \in GS(\psi)$ on Ω_n given $\eta \in \bar{\Omega}_n$.

b) *Each $\mu \in \mathcal{M}$ that satisfies (5.7) belongs to $GS(\psi)$.*

c) *There is a constant C_ψ such that for each $\mu \in GS(\psi)$ and each $\omega \in \Omega$*

$$\exp(-C_\psi \, n^{d-1}) \le \frac{\mu(A_n(\omega))}{e^{\psi_n(\omega) - P_{n,1}(\psi)}} \le \exp(C_\psi \, n^{d-1}). \tag{5.8}$$

(These inequalities are equivalent to $|I_{\alpha^{\Lambda_n}} + \psi_n - P_{n,1}(\psi)| \le C_\psi \, n^{d-1}$.)

5.2.5 Remark In statistical mechanics, (5.7) is often taken as the defining property of Gibbs measures, see e.g. the monograph of Georgii [23]. In an ergodic theoretic context, however, the definition we use seems preferable, because it can be rather flexibly extended to more general dynamical systems, see e.g., [53], [28]. ◇

Before we prove the theorem, we note two corollaries:

5.2.6 Corollary *In dimension $d = 1$ each regular local energy function ψ possesses at most one stationary (i.e., shift-invariant) ergodic Gibbs measure. (The existence of such a measure is a corollary to the main result of the next section.)*

Proof: Let $\mu, \mu' \in GS(\psi, \mathcal{T})$, μ ergodic. Because of (5.8), μ and μ' are mutually absolutely continuous so that $\mu = \mu'$ by Lemma 2.2.2. □

5.2.7 Corollary *If ψ is a regular local energy function, then*

$$GS(\psi, \mathcal{T}) \subseteq ES(\psi, \mathcal{T}) .$$

Proof: Let $\mu \in GS(\psi, \mathcal{T})$. From estimate (5.8) we obtain for each $n > 0$

$$H_\mu(\alpha^{\Lambda_n}) + \int \psi_n \, d\mu = \int (I_{\alpha^{\Lambda_n}} + \psi_n) \, d\mu \geq P_{n,1}(\psi) - C_\psi \, n^{d-1} .$$

Therefore

$$h(\mu) + \int \psi \, d\mu = \lim_{n\to\infty} \frac{1}{\lambda_n} \left(H_\mu(\alpha^{\Lambda_n}) + \int \psi_n \, d\mu \right) \geq \limsup_{n\to\infty} \frac{1}{\lambda_n} P_{n,1}(\psi) = p_1(\psi).$$

As we have seen in Example 4.4.3 that $p_\delta(\psi) = p_1(\psi)$ for all $\delta \in (0, 1)$, it follows from Theorem 4.4.11 that $p(\psi) = p_1(\psi)$. Hence $h(\mu) + \int \psi \, d\mu = p(\psi)$. □

Proof of Theorem 5.2.4:

a) Let $\mu \in GS(\psi)$. We start by proving that

$$\mu([z] \mid \mathcal{H}_n)(\omega) = \exp(\Psi_{\tau_z}(\underline{z}, \eta)) \cdot \mu([\underline{z}] \mid \mathcal{H}_n)(\omega) \tag{5.9}$$

for all $z \in \Omega_n$ and and μ-a.e. $\omega = (z', \eta) \in \Omega$. To this end fix $z \in \Omega_n$ and pick an arbitrary set $H \in \mathcal{H}_n$. By definition of $\mathcal{H}_n$ we have $\tau^{-1} H = H$ for all $\tau \in \mathcal{E}_n$. Therefore

$$\begin{aligned} \mu(H \cap [z]) &= \mu(\tau_z^{-1}(H \cap [\underline{z}])) = \int_{H \cap [\underline{z}]} e^{\Psi_{\tau_z}(z', \eta)} \, d\mu(z', \eta) \\ &= \int_{H \cap [\underline{z}]} e^{\Psi_{\tau_z}(\underline{z}, \eta)} \, d\mu(z', \eta) \\ &= \int_H e^{\Psi_{\tau_z}(\underline{z}, \eta)} \cdot \mu([\underline{z}] \mid \mathcal{H}_n)(z', \eta) \, d\mu(z', \eta) . \end{aligned}$$

This proves (5.9). Summing up that identity over all $z \in \Omega_n$ it follows that

$$1 = \mu(\Omega \mid \mathcal{H}_n)(\omega) = \left(\sum_{z \in \Omega_n} \exp \Psi_{\tau_z}(\underline{z}, \eta) \right) \cdot \mu([\underline{z}] \mid \mathcal{H}_n)(\omega)$$

for μ-a.e. $\omega \in \Omega$. Together with (5.9) this yields (5.7).

b) Let $\tilde{\tau} \in \mathcal{E}$ and consider n so large that $\tilde{\tau} \in \mathcal{E}_n$. Let $z \in \Omega_n$ and $\tilde{\tau}^{-1}[z] =: [\tilde{z}]$. Then $\tau_{\tilde{z}}^{-1}(\underline{z}, \eta) = (\tilde{z}, \eta) = \tilde{\tau}^{-1}(z, \eta)$ and $\tau_z^{-1}(\underline{z}, \eta) = (z, \eta)$ so that

$$(\psi_n \circ \tau_{\tilde{z}}^{-1} - \psi_n \circ \tau_z^{-1})(\underline{z}, \eta) = \psi_n(\tilde{\tau}^{-1}(z, \eta)) - \psi_n(z, \eta) = \Psi_{\tilde{\tau}}^n(z, \eta) .$$

It follows that for $\eta \in \bar{\Omega}_n$

$$\begin{aligned}
\frac{Q_n(\tilde{z}, \eta)}{Q_n(z, \eta)} &= \exp(\Psi_{\tau_{\tilde{z}}}(\underline{z}, \eta) - \Psi_{\tau_z}(\underline{z}, \eta)) \\
&= \lim_{n \to \infty} \exp(\psi_n \circ \tau_{\tilde{z}}^{-1} - \psi_n \circ \tau_z^{-1})(\underline{z}, \eta) \\
&= \lim_{n \to \infty} \exp(\Psi_{\tilde{\tau}}^n(z, \eta)) \\
&= \exp(\Psi_{\tilde{\tau}}(z, \eta)) .
\end{aligned}$$

Suppose now that $\mu \in \mathcal{M}$ satisfies (5.7). Then

$$\begin{aligned}
\mu(\tilde{\tau}^{-1}[z]) &= \mu([\tilde{z}]) = \int Q_n(\tilde{z}, \eta) \, d\mu(z', \eta) \\
&= \int \exp(\Psi_{\tilde{\tau}}(z, \eta)) Q_n(z, \eta) \, d\mu(z', \eta) \\
&= \int \exp(\Psi_{\tilde{\tau}}(z, \eta)) \mu([z] \mid \mathcal{H}_n)(z', \eta) \, d\mu(z', \eta) \\
&= \int_{[z]} \exp(\Psi_{\tilde{\tau}}(z, \eta)) \, d\mu(z', \eta) \\
&= \int_{[z]} \exp(\Psi_{\tilde{\tau}}(\omega)) \, d\mu(\omega) .
\end{aligned}$$

As the family of cylinder sets $[z]$, $z \in \Omega_n$, $n > 0$, is an $\cap$-stable generator of the Borel σ-algebra $\mathcal{B}$ of Ω, it follows that $\tilde{\tau}\mu = e^{\Psi_{\tilde{\tau}}} \mu$ (see A.4.7). As this is true for all $\tilde{\tau} \in \mathcal{E}$, we conclude that $\mu \in GS(\psi)$.

c) Because of (5.6) there is a constant $C_0 > 0$ such that, using the above notation,

$$\exp(-C_0 \, n^{d-1}) \leq \frac{\sum_{z' \in \Omega_n} \exp \Psi_{\tau_{z'}}^n(\underline{z}, \eta)}{\exp \Psi_{\tau_z}^n(\underline{z}, \eta)} \cdot Q_n(z, \eta) \leq \exp(C_0 \, n^{d-1})$$

for all $n > 0$ and $(z, \eta) \in \Omega_n \times \bar{\Omega}_n$. As

$$\frac{\exp \Psi_{\tau_{z'}}^n(\underline{z}, \eta)}{\exp \Psi_{\tau_z}^n(\underline{z}, \eta)} = \frac{\exp \psi_n(z', \eta) \cdot \exp(-\psi_n(\underline{z}, \eta))}{\exp \psi_n(z, \eta) \cdot \exp(-\psi_n(\underline{z}, \eta))} = \frac{\exp \psi_n(z', \eta)}{\exp \psi_n(z, \eta)} ,$$

it follows that

$$\exp(-C_0 \, n^{d-1}) \leq \frac{\sum_{z' \in \Omega_n} \exp \psi_n(z', \eta)}{\exp \psi_n(z, \eta)} \cdot Q_n(z, \eta) \leq \exp(C_0 \, n^{d-1}) . \tag{5.10}$$

Because of (5.4) and (5.3) in Lemma 5.1.4, there is a constant $C_1 > 0$ such that

$$\exp(-C_1\, n^{d-1}) \leq \frac{\sum_{z' \in \Omega_n} \exp \psi_n(z', \eta)}{\exp P_{n,1}(\psi)} \leq \exp(C_1\, n^{d-1})$$

and such that for each $\eta' \in \bar{\Omega}_n$

$$\exp(-C_1\, n^{d-1}) \leq \frac{\exp \psi_n(z, \eta)}{\exp \psi_n(z, \eta')} \leq \exp(C_1\, n^{d-1}) .$$

Hence there is a constant $C_\psi := C_0 + 2C_1 > 0$ such that

$$\exp(-C_\psi\, n^{d-1}) \leq \frac{Q_n(z, \eta)}{\exp(\psi_n(z, \eta') - P_{n,1}(\psi))} \leq \exp(C_\psi\, n^{d-1}) . \tag{5.11}$$

Observing that $Q_n(z, \eta) = \mu([z] \mid \mathcal{H}_n)(z', \eta)$ by (5.7), it suffices to integrate (5.11) over $\omega = (z', \eta)$ to obtain (5.8). □

5.2.8 Exercise Let $\Omega = \Sigma^{\mathbb{Z}}$ be a shift space and denote by $\mathcal{T}$ the canonical action on Ω. Let ψ be a regular local energy function and fix some $\mu \in GS(\psi, \mathcal{T})$. Prove that the two-sided tail-field $\mathcal{H}_\infty := \bigcap_{n \in \mathbb{N}} \mathcal{H}_n$ is μ-trivial, i.e., $\mu(H) = 0$ or $\mu(H) = 1$ for all $H \in \mathcal{H}_\infty$ (cf. Exercise 3.2.25). *Hint:* Use (5.11) to show that there is a constant $C > 0$ such that $\mu(H \cap [z]) \geq C \cdot \mu(H) \cdot \mu([z])$ for all $H \in \mathcal{H}_\infty$, $z \in \Omega_n$ and $n \in \mathbb{N}$.

Comment: From this together with Exercise 3.2.25 it follows that $(\Omega, \mathcal{B}, \mu, \mathcal{T})$ is mixing and that $h(\mu, \alpha, \mathcal{T}) > 0$ for each nontrivial μ-partition of Ω. ◇

5.3 Equilibrium states are Gibbs measures

The inclusion $GS(\psi, \mathcal{T}) \subseteq ES(\psi, \mathcal{T})$ for regular local energy functions ψ was the main result of the previous section. Here we prove the converse and obtain

5.3.1 Theorem (Equilibrium states = Gibbs measures)
Let ψ be a regular local energy function. Then $ES(\psi, \mathcal{T}) = GS(\psi, \mathcal{T})$.

5.3.2 Corollary *If $G = \mathbb{Z}$, then there exists exactly one Gibbs measure for each regular local energy function ψ. This Gibbs measure is also the unique equilibrium state for ψ (and is hence ergodic).*

Proof: As $ES(\psi, \mathcal{T}) = GS(\psi, \mathcal{T})$, the existence follows from Corollary 4.2.5 and the example thereafter. If there were more than one equilibrium state, Theorem 4.3.9 would guarantee the existence of at least two *ergodic* equilibrium states. But this is excluded by Corollary 5.2.6. □

5.3.3 Corollary *If ψ is a regular local energy function, then*

$$p(\psi) = \lim_{n\to\infty} \frac{1}{\lambda_n} P_{n,1}(\psi) .$$

(It is not just the lim sup*!)*

Proof: As $GS(\psi, \mathcal{T}) = ES(\psi, \mathcal{T}) \neq \emptyset$, there is $\mu \in GS(\psi, \mathcal{T})$. Because of (5.8) in Theorem 5.2.4,

$$-\frac{C_\psi}{n} \leq \frac{1}{\lambda_n} \left(P_{n,1}(\psi) - (\psi_n(\omega) - \log \mu(A_n(\omega)))\right) \leq \frac{C_\psi}{n}$$

for each $\omega \in \Omega$. Integrating these inequalities yields

$$-\frac{C_\psi}{n} \leq \frac{1}{\lambda_n} P_{n,1}(\psi) - \left(\mu(\psi) + \frac{1}{\lambda_n} H_\mu(\alpha^{\Lambda_n})\right) \leq \frac{C_\psi}{n} .$$

In the limit $n \to \infty$ this gives $\lim_{n\to\infty} \frac{1}{\lambda_n} P_{n,1}(\psi) = \mu(\psi) + h(\mu) = p(\psi)$. □

Proof of Theorem 5.3.1: As the shift action $\mathcal{T}$ is expansive with constant 1 (Example 4.5.3), as the set D of regular ϕ is dense in $C(X)$ (Remark 5.1.2-b) and as all $\phi, \psi \in D$ satisfy (4.31) (Lemma 5.1.4), it follows from Theorem 4.5.9 and the remark thereafter that $ES(\psi, \mathcal{T})$ is the closed convex hull of $\bigcup_{\phi\in D} CES_\phi^\xi(1, \psi, \mathcal{T})$ for each $\xi \in \Omega$. By Lemma 5.2.2 the set $GS(\psi, \mathcal{T})$ is closed and convex. Hence it suffices to prove for some $\xi \in \Omega$ that $CES_\phi^\xi(1, \psi, \mathcal{T}) \subseteq GS(\psi, \mathcal{T})$ for each regular ϕ.

Let ϕ be regular. By Remarks 4.5.8 and 4.5.10, $CES_\phi^\xi(1, \psi, \mathcal{T})$ is the set of those constructible 1-equilibrium states whose construction is based on weight functions $\psi^{(n)} = \psi_n + t_n\phi_n$ where $(t_n)_n$ is any null sequence in $[-1, 1]$, and on the sets $E_n^\xi = \{\omega \in \Omega : \omega_g = \xi_g \ \forall g \in G \setminus \Lambda_n\}$ where $\xi \in \Omega$ is fixed. These sets have the property that

$$\tau E_n^\xi = E_n^\xi \quad \text{for all } \tau \in \mathcal{E}_n. \tag{5.12}$$

So let $\tau \in \mathcal{E}_m$ for some $m > 0$. Recall from Lemma 5.1.6 that $\lim_{n\to\infty} \|\Psi_\tau - \Psi_\tau^n\|_\infty = 0$ where $\Psi_\tau^n = \psi_n \circ \tau^{-1} - \psi_n$. Similarly there exists Φ_τ such that $\lim_{n\to\infty} \|\Phi_\tau - \Phi_\tau^n\|_\infty = 0$ where $\Phi_\tau^n = \phi_n \circ \tau^{-1} - \phi_n$. Let $\Psi_\tau^{(n)} = \psi^{(n)} \circ \tau^{-1} - \psi^{(n)}$. Then

$$\begin{aligned}
\limsup_{n\to\infty} \|\Psi_\tau - \Psi_\tau^{(n)}\|_\infty &= \limsup_{n\to\infty} \|\Psi_\tau^n - \Psi_\tau^{(n)}\|_\infty = \limsup_{n\to\infty} \|t_n \Phi_\tau^n\|_\infty \\
&\leq \limsup_{n\to\infty} |t_n| \cdot \|\Phi_\tau\|_\infty = 0 .
\end{aligned} \tag{5.13}$$

For $g \in G$ let

$$\tau_{[g]} := T^{-g} \circ \tau \circ T^g . \tag{5.14}$$

Then $\tau_{[g]}^{-1} = (\tau^{-1})_{[g]}$ and $(\tau_{[g]})_{[h]} = \tau_{[g+h]}$ for all $g, h \in G$. If $g \in \Lambda_l$, then $\tau_{[g]} \in \mathcal{E}_{m+l-1}$. Later we prove that

$$\lim_{n\to\infty} \sup_{g\in\Lambda_{l(n)}} \|\Psi_\tau^{(n)} \circ T^g - \Psi_{\tau_{[g]}}^{(n)}\|_\infty = 0 \tag{5.15}$$

when $l = l(n)$ is chosen in the following way:

$$l = l(n) = \left[n \cdot \left(1 - \frac{1}{p_n}\right)\right] \qquad \text{(in particular: } l \le (p_n - 1)(n - l)\text{)}$$

where $p_n \to \infty$ and $n - l \geq \frac{n}{p_n} \to \infty$ in such a way that

$$\lim_{n\to\infty} p_n^{d-1} \sum_{j=n-l-m}^{\infty} (j+m)^{d-1}(\delta_j(\psi) + \delta_j(\phi)) = 0\,. \tag{5.16}$$

Since ψ and ϕ are regular, such a choice is always possible. Observe also that

$$\lim_{n\to\infty} \frac{l(n)}{n} = \lim_{n\to\infty} \frac{p_n - 1}{p_n} = 1 \quad \text{and hence} \quad \lim_{n\to\infty} \frac{\lambda_{l(n)}}{\lambda_n} = 1\,.$$

Let $\mu \in CES_\phi^\xi(1, \psi, \mathcal{T})$. Then

$$\mu = \lim_{i\to\infty} \frac{1}{\lambda_{n_i}} \sum_{g\in\Lambda_{n_i}} T^g \pi_{n_i} = \lim_{i\to\infty} \frac{1}{\lambda_{l_i}} \sum_{g\in\Lambda_{l_i}} T^g \pi_{n_i}$$

for some subsequence (n_i) and $l_i := l(n_i)$, where the measures

$$\pi_n = \sum_{z\in E_n^\xi} \delta_z \cdot \frac{\exp \psi^{(n)}(z)}{\exp P(\psi^{(n)}, E_n^\xi)}$$

are defined as in (4.12).

Recall that $\tau \in \mathcal{E}_m$. Let n be so large that $n > m + l$, $l = l(n)$, and consider $g \in \Lambda_l$ and $f \in C(X)$. Then $\tau_{[g]} \in \mathcal{E}_{l+m-1} \subseteq \mathcal{E}_n$ so that $\tau_{[g]} E_n^\xi = E_n^\xi$ and

$$\begin{aligned}
&(T^g \pi_n)(f \circ \tau) \\
&= \pi_n(f \circ \tau \circ T^g) = \pi_n(f \circ T^g \circ \tau_{[g]}) \quad \text{because of (5.14)} \\
&= \sum_{z\in E_n^\xi} (f \circ T^g)(\tau_{[g]}(z)) \cdot \exp\left(\psi^{(n)}(z) - P(\psi^{(n)}, E_n^\xi)\right) \\
&= \sum_{z\in \tau_{[g]} E_n^\xi} (f \circ T^g)(z) \cdot \exp\left(\psi^{(n)}(\tau_{[g]}^{-1}(z)) - P(\psi^{(n)}, E_n^\xi)\right) \\
&= \sum_{z\in E_n^\xi} (f \circ T^g)(z) \cdot \exp\left(\psi^{(n)}(z) - P(\psi^{(n)}, E_n^\xi)\right) \cdot \exp\left(\psi^{(n)}(\tau_{[g]}^{-1}(z)) - \psi^{(n)}(z)\right) \\
&= \pi_n\left((f \circ T^g) \cdot \exp\left(\Psi_{\tau_{[g]}}^{(n)}\right)\right) \\
&= \pi_n\left((f \cdot e^{\Psi_\tau^{(n)}}) \circ T^g \cdot \exp\left(\Psi_{\tau_{[g]}}^{(n)} - \Psi_\tau^{(n)} \circ T^g\right)\right)\,,
\end{aligned}$$

so that

$$\begin{aligned}
\mu(f\circ\tau) &= \lim_{i\to\infty}\frac{1}{\lambda_{l_i}}\sum_{g\in\Lambda_{l_i}} T^g\pi_{n_i}(f\circ\tau)\\
&= \lim_{i\to\infty}\frac{1}{\lambda_{l_i}}\sum_{g\in\Lambda_{l_i}} \pi_{n_i}\left((f\cdot e^{\Psi_\tau^{(n_i)}})\circ T^g\cdot\exp(\Psi_{\tau_{[g]}}^{(n_i)}-\Psi_\tau^{(n_i)}\circ T^g)\right),\\
&= \lim_{i\to\infty}\frac{1}{\lambda_{l_i}}\sum_{g\in\Lambda_{l_i}} \pi_{n_i}\left((f\cdot e^{\Psi_\tau})\circ T^g\right) \quad \text{because of (5.13) and (5.15),}\\
&= \mu(f\cdot e^{\Psi_\tau}).
\end{aligned}$$

As this is true for all $f\in C(X)$ and $\tau\in\mathcal{E}$ it follows that $\tau\mu=\mu\cdot e^{\psi_\tau}$ for all $\tau\in\mathcal{E}$, i.e., $\mu\in GS(\psi,\mathcal{T})$.

We still have to prove (5.15): Let $\tilde\psi := \psi+t_n\phi$ so that $\tilde\psi_n=\psi^{(n)}$. (We suppress the dependence on n via t_n in this notation.) Then, for $g\in\Lambda_l$,

$$\begin{aligned}
&\|\Psi_\tau^{(n)}\circ T^g-\Psi_{\tau_{[g]}}^{(n)}\|_\infty\\
&= \|\psi^{(n)}\circ\tau^{-1}\circ T^g-\psi^{(n)}\circ T^g-\psi^{(n)}\circ\tau_{[g]}^{-1}+\psi^{(n)}\|_\infty\\
&= \|(\psi^{(n)}\circ T^g-\psi^{(n)})\circ\tau_{[g]}^{-1}-(\psi^{(n)}\circ T^g-\psi^{(n)})\|_\infty \quad \text{in view of (5.14)}\\
&= \left\|\left(\sum_{k\in(\Lambda_n+g)\setminus\Lambda_n}\tilde\psi\circ T^k-\sum_{k\in\Lambda_n\setminus(\Lambda_n+g)}\tilde\psi\circ T^k\right)\circ\tau_{[g]}^{-1}\right.\\
&\qquad\left.-\left(\sum_{k\in(\Lambda_n+g)\setminus\Lambda_n}\tilde\psi\circ T^k-\sum_{k\in\Lambda_n\setminus(\Lambda_n+g)}\tilde\psi\circ T^k\right)\right\|_\infty\\
&= \left\|\sum_{k\in(\Lambda_n+g)\setminus\Lambda_n}\left(\tilde\psi\circ T^k\circ\tau_{[g]}^{-1}-\tilde\psi\circ T^k\right)\right.\\
&\qquad\left.-\sum_{k\in\Lambda_n\setminus(\Lambda_n+g)}\left(\tilde\psi\circ T^k\circ\tau_{[g]}^{-1}-\tilde\psi\circ T^k\right)\right\|_\infty\\
&\le \sum_{k\in(\Lambda_n+g)\setminus\Lambda_n}\|(\tilde\psi\circ\tau_{[g-k]}^{-1}-\tilde\psi)\circ T^k\|_\infty\\
&\qquad+\sum_{k\in\Lambda_n\setminus(\Lambda_n+g)}\|(\tilde\psi\circ\tau_{[g-k]}^{-1}-\tilde\psi)\circ T^k\|_\infty
\end{aligned}$$

in view of (5.14). By Remark 5.1.5-b we have $(\tau_{[g-k]}^{-1}\omega)_i=(T^{g-k}\omega)_{i-(g-k)}=\omega_i$ for all indices i with $|i|\le|g-k|-m$, in particular if $|i|\le|k|-l-m$. Therefore

$$\|\tilde\psi\circ\tau_{[g-k]}^{-1}-\tilde\psi\|_\infty\le\delta_{|k-g|-m}(\tilde\psi)\le\delta_{|k|-l-m}(\tilde\psi).$$

As $(\Lambda_n + g) \setminus \Lambda_n = \bigcup_{j=n}^{\infty} (\Lambda_n + g) \cap (\Lambda_{j+1} \setminus \Lambda_j)$, we can estimate the first sum by

$$\sum_{k \in (\Lambda_n + g) \setminus \Lambda_n} \|(\tilde{\psi} \circ \tau_{[g-k]}^{-1} - \tilde{\psi})\|_\infty \leq \sum_{j=n}^{\infty} d \cdot (2n+1)^{d-1} \cdot \delta_{j-l-m}(\tilde{\psi}) .$$

For the second sum decompose $\Lambda_n \setminus (\Lambda_n + g) = \bigcup_{j=n}^{\infty} \Lambda_n \cap ((\Lambda_{j+1} + g) \setminus (\Lambda_j + g))$. As $|k - g| \geq j$ for $k \notin (\Lambda_j + g)$, the second sum can be estimated by

$$\sum_{k \in \Lambda_n \setminus (\Lambda_n + g)} \|(\tilde{\psi} \circ \tau_{[g-k]}^{-1} - \tilde{\psi})\|_\infty \leq \sum_{j=n}^{\infty} d \cdot (2n+1)^{d-1} \cdot \delta_{j-m}(\tilde{\psi}) .$$

Observing that $\delta_{j-m}(\tilde{\psi}) \leq \delta_{j-l-m}(\tilde{\psi})$ we can put these estimates together and obtain

$$\|\Psi_\tau^{(n)} \circ T^g - \Psi_{\tau[g]}^{(n)}\|_\infty \leq 2d \sum_{j=n-l-m}^{\infty} (2n+1)^{d-1} \cdot \delta_j(\tilde{\psi}) .$$

As $n \leq j + m + l \leq j + m + (p_n - 1)(n - l) \leq p_n(j + m)$ for each $j \geq n - l - m$, we conclude that

$$\|\Psi_\tau^{(n)} \circ T^g - \Psi_{\tau[g]}^{(n)}\|_\infty \leq \text{const} \cdot p_n^{d-1} \sum_{j=n-l-m}^{\infty} (j+m)^{d-1} (\delta_j(\psi) + |t_n|\, \delta_j(\phi)) ,$$

and this tends to 0 as $n \to \infty$ in view of (5.16). This finishes the proof of (5.15). □

5.3.4 Remark (Nonuniqueness of equilibrium states) In the literature there are several examples of local energy functions on one-dimensional shift spaces which are not regular and admit in fact at least two equilibrium states.

Hofbauer [29] studies a local energy function ψ on $\Omega = \{0,1\}^{\mathbb{Z}_+}$ defined as follows: Let (a_k) be a sequence of real numbers with $\lim_{k \to \infty} a_k = 0$. Set $s_k := a_0 + \ldots + a_k$. For $k \geq 0$ denote $M_k := \{\omega \in \Omega : \omega_0 = \ldots = \omega_{k-1} = 1,\ \omega_k = 0\}$ and define

$$\psi(\omega) := a_k \quad \text{for } x \in M_k \quad \text{and} \quad \psi(11\ldots) = 0 .$$

Then $\psi : \Omega \to \mathbb{R}$ is continuous, so that there exists at least one equilibrium state for ψ. Hofbauer proves that there is more than one equilibrium state if and only if $\sum_k e^{s_k} = 1$ and $\sum_k k e^{S_k} < \infty$. In that case one of these equilibrium states is the unit mass $\delta_{11\ldots}$. We give a similar example in the context of one-dimensional maps in Example 6.2.7. The function ψ in our example, however, is not in $C(X)$ but just in $USC(X)$, because it takes the value $-\infty$ at one point.

Another example is due to Krieger [35]. He constructs a closed shift-invariant subset X of $\Omega = \{0, \ldots, 3\}^{\mathbb{Z}}$ such that the restriction of the shift action to X

has two measures of maximal entropy although the set X is specified by relatively simple rules on the symbols and although it has a dense orbit under the shift action. In view of Remark 4.2.7 this means that the upper semicontinuous function $\psi^* : X \to [-\infty, \infty)$, $\psi^*(\omega) = 0$ if $\omega \in X$, $\psi^*(\omega) = -\infty$ otherwise, admits two equilibrium states for the action of the full shift.

The nonuniqueness of equilibrium states for the two-dimensional Ising model is shown in Section 5.5.

5.4 Markov chains

Among the most basic examples of equilibrium states and Gibbs measures are stationary measures for Markov chains over a finite alphabet. A *Markov chain* over a finite alphabet Σ is specified by a stochastic $\Sigma \times \Sigma$ matrix $Q = (q_{\sigma\tau})$, i.e.,

$$q_{\sigma\tau} \geq 0 \text{ for all } \sigma, \tau \in \Sigma \text{ and } \sum_{\tau \in \Sigma} q_{\sigma\tau} = 1 \text{ for all } \sigma \in \Sigma.$$

The dynamical model for such a chain is the shift space $(\Omega, \mathcal{T})$ where $\Omega = \Sigma^{\mathbb{Z}}$, $\mathcal{T} = (T^k : k \in \mathbb{Z})$, and T is the left shift on Ω, $(T\omega)_i = \omega_{i+1}$. Recall from Example 4.5.3 that $\mathcal{T}$ is expansive so that the entropy function is upper semicontinuous (Theorem 4.5.6).

With Q we associate the local energy function $\psi \in USC(\Omega)$ defined by $\psi(\omega) = \log q_{\omega_{-1}\omega_0}$. (As usual let $\log 0 = -\infty$.) Note that $\psi_m(\omega) = \sum_{k=-m+1}^{m-1} \log q_{\omega_{k-1}\omega_k}$ depends only on the values ω_i for $i \in \{-m\} \cup \Lambda_m$.

In this situation Theorem 4.2.3 guarantees the existence of an equilibrium state μ for ψ. This equilibrium state need not be unique in general. However, if Q is *strictly positive*, i.e., if $q_{\sigma\tau} > 0$ for all $\sigma, \tau \in \Sigma$, then ψ is a real-valued function of finite range, so a fortiori ψ is regular, and ψ has a unique equilibrium state μ which is at the same time its unique Gibbs measure by Corollary 5.2.6. In the next theorem we identify μ as the unique invariant Markov measure for Q.

5.4.1 Theorem (Markov measures as equilibrium states)
Let Q be strictly positive and let ψ be as above. Then ψ has a unique equilibrium state which is at the same time the unique invariant Markov measure for Q, i.e., for $z \in \Omega_n$

$$\mu([z] \mid \mathcal{H}_n^-)(\omega) = q_{\omega_{-n} z_{-n+1}} \prod_{i=-n+2}^{n-1} q_{z_{i-1} z_i}$$

where $\mathcal{H}_n^-$ is the σ-algebra generated by the coordinate mappings $\omega \mapsto \omega_i$, $i \leq -n$. Furthermore, $p(\psi) = 0$. (For the notation Ω_n and for a comparison of $\mathcal{H}_n^-$ with $\mathcal{H}_n$ see Section 5.2.)

Proof: Existence and uniqueness of the equilibrium state μ follow from Corollary 5.3.2. In particular μ is ergodic. For notational convenience we write $q(\sigma, \tau)$ instead of $q_{\sigma\tau}$. It follows from Theorem 5.2.4 that

$$e^{-C_\psi} \leq \frac{\mu([z])\, e^{P_{n,1}(\psi)}}{q(\eta_{-n}, z_{-n+1}) \cdot \prod_{i=-n+2}^{n-1} q(z_{i-1}, z_i)} \leq e^{C_\psi} \tag{5.17}$$

for each $(z, \eta) \in \Omega_n \times \bar{\Omega}_n$. Summing these inequalities over $z \in \Omega_n$ yields after some rearrangement

$$e^{-C_\psi}\, e^{P_{n,1}(\psi)} \leq \sum_{z \in \Omega_n} \left(q(\eta_{-n}, z_{-n+1}) \prod_{i=-n+2}^{n-1} q(z_{i-1}, z_i) \right) \leq e^{C_\psi}\, e^{P_{n,1}(\psi)}$$

where the sum evaluates to $\sum_{z_{n-1} \in \Sigma} q^{(2n-1)}(\eta_{-n}, z_{n-1}) = 1$ because $Q^{(2n-1)}$ is also a stochastic matrix. Hence $|\log P_{n,1}(\psi)| \leq C_\psi$ so that $p(\psi) = p_1(\psi) = 0$.

Let now ν be a stationary Markov measure for Q, i.e.,

$$\nu([z]) = \pi_{z_{-n+1}} \cdot \prod_{i=-n+2}^{n-1} q(z_{i-1}, z_i)$$

for some probability vector π on Σ. In view of (5.17)

$$\frac{\nu([z])}{\mu([z])} \leq e^{C_\psi} \frac{\pi_{z_{-n+1}}}{q(\eta_{-n}, z_{-n+1})} e^{P_{n,1}(\psi)} \leq e^{2C_\psi} \sup_{\sigma,\tau \in \Sigma} \frac{\pi_\tau}{q(\sigma, \tau)} < \infty$$

for all $z \in \Omega_n$ and uniformly in n. Hence $\nu \ll \mu$, and as μ is ergodic, it follows that $\nu = \mu$. □

5.4.2 Exercise Let π be the one-dimensional marginal distribution of the unique stationary Markov measure μ for the strictly positive matrix $(q_{\sigma\tau})_{\sigma,\tau \in \Sigma}$ as in the proof of the theorem. Denote by T the shift on Ω. Show that

$$h(\mu) = -\sum_{\sigma,\tau \in \Sigma} \pi_\sigma\, q_{\sigma\tau} \log q_{\sigma\tau}\,.$$

5.5 Equilibrium states of the Ising model

Historically one of the first models from statistical mechanics that was attacked by rigorous mathematical methods was the Ising model. In Example 1.2.2 we described its basic ingredients: the alphabet $\Sigma = \{-1, +1\}$ representing "spin up" and "spin down" at the sites of a "lattice gas" on $G = \mathbb{Z}^d$, the configuration space $\Omega = \Sigma^G$ equipped with the shift action $\mathcal{T}$, and the local energy function $\psi(\omega) = -\sum_{j=1}^{d}(\omega_0\omega_{e_j} + \omega_0\omega_{-e_j})$ describing the interaction between neighbouring spins. Here we consider a further contribution to the local energy, namely the additional term $-B\phi(\omega) := -B\omega_0$ that describes the effect of a magnetic field B on the spin at site 0. So we are led to study the local energy

$$\psi_{\beta,B}(\omega) := -\beta\cdot(\psi - B\phi)(\omega) = -\beta\cdot(\psi(\omega) - B\omega_0)$$

where $\beta \geq 0$ denotes the inverse temperature, see Remark 1.1.6. As $\psi_{\beta,B}$ has finite range, it is regular, and therefore $ES(\psi_{\beta,B}, \mathcal{T}) = GS(\psi_{\beta,B}, \mathcal{T})$ is nonempty. The question is, whether the equilibrium states are unique. It turns out that the case $d = 1$ is fundamentally different from the cases $d \geq 2$:

5.5.1 Theorem (Phase transitions in the Ising model)

a) *If $d = 1$, then the Ising model has exactly one equilibrium state for each $\beta > 0$ and $B \in \mathbb{R}$.*

b) *If $d \geq 2$ and $\epsilon > 0$, then there exists $\beta_d(\epsilon) > 0$ such that for $\beta > \beta_d(\epsilon)$ the picture, in dependence on the magnetic field B, looks as follows:*
$B > 0$: If $\mu \in ES(\psi_{\beta,B}, \mathcal{T})$, then $\mu\{\omega_0 = +1\} \geq 1 - \epsilon$.
$B < 0$: If $\mu \in ES(\psi_{\beta,B}, \mathcal{T})$, then $\mu\{\omega_0 = -1\} \geq 1 - \epsilon$.
$B = 0$: There are $\mu^+, \mu^- \in ES(\psi_{\beta,0}, \mathcal{T})$ such that

$$\mu^+\{\omega_0 = +1\} = \mu^-\{\omega_0 = -1\} \geq 1 - \epsilon\,.$$

Furthermore, μ^+ is the weak limit of a sequence $\mu_i^+ \in ES(\psi_{\beta,B_i}, \mathcal{T})$ where $B_i \searrow 0$, and μ^- is the weak limit of a sequence $\mu_i^- \in ES(\psi_{\beta,B_i}, \mathcal{T})$ where $B_i \nearrow 0$.

5.5.2 Remark For high β (i.e., low temperature) and $d \geq 2$ this result can be interpreted as *spontaneous magnetization* of the Ising ferromagnet: When the magnetic field B is positive, the "average magnetization per spin" $\mu(\phi) = \int \omega_0\, d\mu(\omega)$ is close to $+1$, and it is close to -1, when $B < 0$. If the magnetic field is turned down and finally switched off, the magnetization keeps at least the level $\pm(1-\epsilon)$, the sign being the same as before the field was turned off. If the field B is driven from positive to negative values, the average magnetization changes abruptly at $B = 0$ from $\mu(\phi) > 1 - 2\epsilon$ to $\mu(\phi) < -1 + 2\epsilon$. At $B = 0$ two equilibrium states can coexist, one favouring "spin up", the other one favouring "spin down". This phenomenon is called a *phase transition*.

For the two-dimensional Ising model the critical inverse temperature β_c above which a phase transition occurs is known to be $\beta_c = \frac{1}{2}\sinh^{-1}(1) \approx 0.4407$. The story that led to this number and its deeper understanding is told in [17, Section 8]. A typical configuration at $\beta = 0.4410$ is shown in Figure 5.1. For a discussion of this figure in the light of large deviations theory see also Remark 5.6.2. The problem of how to generate such a picture is out of the scope of this text. A reliable and fast method is to use the so called *Swendsen–Wang algorithm.* ◇

5.5.3 Exercise Show that if $B_1 < B_2$ and $\mu_i \in ES(\psi_{\beta,B_i}, T)$ $(i = 1, 2)$, then $\mu_1(\phi) \leq \mu_2(\phi)$. ◇

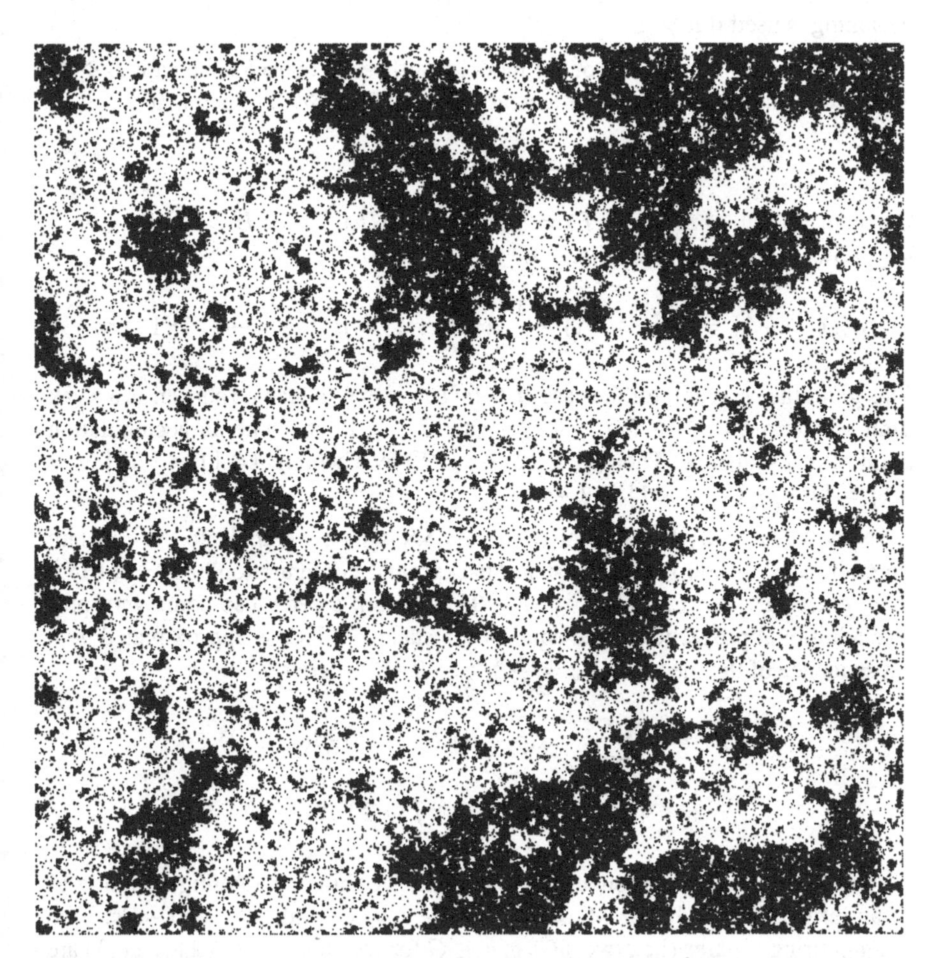

Figure 5.1: Typical Ising configurations on a 800 × 800 grid. $\beta = 0.4410$ is slightly larger than the critical parameter $\beta_c = \frac{1}{2}\sinh^{-1}(1)$; so one kind of spins prevails. The "ocean–island–pond–etc." structure discussed in Remark 5.6.2 is clearly visible.

Proof of the theorem: Recall first that for shift systems the entropy function is upper semicontinuous and that $p_1(\psi) = \sup_{\delta>0} p_\delta(\psi)$ for all $\psi \in USC(X)$ so that practically all results proved so far are applicable.

a) In the case $d = 1$ the uniqueness of the equilibrium state is an immediate consequence of Corollary 5.3.2.

b) In the case $d \geq 2$ an argument originally due to Peierls [44], and later on made rigorous by Griffiths [26] for the case $d = 2$, shows that for large β and $B = 0$ there exist at least two equilibrium states. We present a version of this argument restricting ourselves to the case $d = 2$. A geometrically more subtle but similar reasoning is used if $d > 2$.

Because of the symmetry of the model it suffices to consider the case $B \geq 0$. Below we prove the existence of a measure $\mu^+ \in ES(-\beta\psi, \mathcal{T})$ such that

$$\mu^+\{\omega_0 = +1\} \geq 1 - \epsilon. \tag{5.18}$$

Consider now $\mu \in ES(\psi_{\beta,B}, \mathcal{T}) = ES(-\beta\psi + \beta B\phi, \mathcal{T})$ for $B > 0$. By Corollary 4.3.6, $\mu(\phi) \geq \mu^+(\phi)$ so that

$$\mu\{\omega_0 = +1\} = \frac{1}{2}(\mu(\phi) + 1) \geq \frac{1}{2}(\mu^+(\phi) + 1) = \mu^+\{\omega_0 = +1\} \geq 1 - \epsilon.$$

This gives the assertion for the case $B > 0$. To see that μ^+ can be chosen as a weak limit of measures $\mu_i^+ \in ES(\psi_{\beta,B_i}, \mathcal{T})$ with $B_i \searrow 0$, let $\tilde{\mu}^+$ be an accumulation point of this sequence. Then $\tilde{\mu}^+ \in ES(-\beta\psi, \mathcal{T})$ by Theorem 4.2.11 and $\tilde{\mu}^+\{\omega_0 = +1\} = \lim_k \mu_{i_k}\{\omega_0 = +1\} \geq 1 - \epsilon$.

Now let $B = 0$. Because of Remark 4.5.8, which we apply to the constant function $\phi = 0$ and to the constant "background configuration" $\xi_+ = (+1)^G$, there exists $\mu^+ \in CES_0^{\xi_+}(1, -\beta\psi, \mathcal{T})$. This remark asserts also some very precise additional properties of the equilibrium state μ^+. That is approximated by a subsequence of the measures $\tilde{\mu}_n = \frac{1}{\lambda_n}\sum_{g \in \Lambda_n} T^g \tilde{\pi}_n$ where

$$\begin{aligned} \tilde{\pi}_n &:= \sum_{z \in E_n^+} \delta_z \cdot \exp(-\beta\psi_n(z) - P(-\beta\psi_n, E_n^+)), \\ E_n^+ &:= \{\omega \in \Omega : \omega_g = +1 \ \forall g \in G \setminus \Lambda_n\}. \end{aligned}$$

For the following combinatorial arguments we think of a configuration ω as of an infinite checkerboard whose squares are covered with $+1$'s and -1's. Two neighbouring squares (i.e., two sites $g, g' \in G$ for which $g - g' \in \{\pm e_1, \pm e_2\}$) are separated by a line segment, their common boundary. Formally denote the set of all such segments by

$$L := \{\{g, g'\} : g' - g \in \{e_1, e_2\}\}.$$

Just as on a real checkerboard, adjacent segments can be connected to a path, and a closed, simply connected path will be denoted by the letter C. Because of the immediate geometric appeal of these notions we dispense with a formally satisfying definition. For our purposes it is sufficient to think of a curve C as of a particular subset of L. As a closed simply connected curve has a connected interior domain (Jordan's curve theorem), we may speak about the interior $I(C)$ of such a curve C and associate to C a map $\tau_C \in \mathcal{E}$ defined by

$$(\tau_C\omega)_i = \begin{cases} -\omega_i & (i \in I(C)), \\ \omega_i & (i \notin I(C)). \end{cases}$$

For a given "square" $g \in G$ let $\mathcal{C}(g)$ denote the set of all closed simply connected curves C with $g \in I(C)$.

Consider $\omega \in E_n^+$ and $g \in \Lambda_n$ such that $\omega_g = -1$. Denote by $A(\omega, g)$ the "connected component" of (-1)-squares which contains g. (For the notion of "connected component" we appeal once more to the intuitive picture of a checkerboard.) $A(\omega, g)$ is bordered by a simply connected closed curve $C = C(\omega, g)$. (This means that C is that part of the boundary of $A(\omega, g)$ which separates g from infinity.) With respect to ω it has the property that $\omega_g\omega_{g'} = -1$ for all $\{g, g'\} \in C$, i.e., each segment of C separates a $(+1)$-square from a (-1)-square. Such a curve C is called a *contour* for ω. Figure 5.2 illustrates these notions. Given $g \in \Lambda_n$ and a simply connected closed curve C we denote by $\Omega_{g,C}$ the set of all $\omega \in \Omega$ for which $\omega_g = -1$ and for which $A(\omega, g)$ is bordered by C. As $\omega \in E_n^+$, the component $A(\omega, g)$ is contained in Λ_n. In particular, if $\{g, g'\} \in C$, then at least one of g and g' belongs to Λ_n.

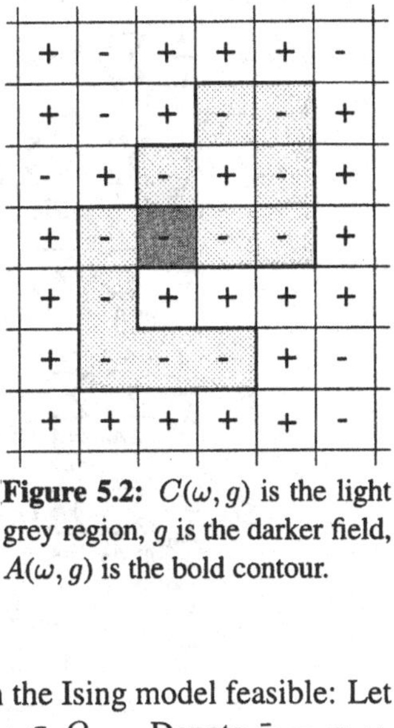

+	-	+	+	+	-
+	-	+	-	-	+
-	+	-	+	-	+
+	-	-	-	-	+
+	-	+	+	+	+
+	-	-	-	+	-
+	+	+	+	+	-

Figure 5.2: $C(\omega, g)$ is the light grey region, g is the darker field, $A(\omega, g)$ is the bold contour.

The following observation makes calculations in the Ising model feasible: Let g and $C = C(\omega, g)$ be as above and suppose that $\omega \in \Omega_{g,C}$. Denote $\bar{\omega} = \tau_C\omega$. Then

$$\begin{aligned} -\beta(\psi_n(\omega) - \psi_n(\tau_C\omega)) &= \beta \sum_{h\in\Lambda_n} \sum_{j\in\{\pm e_1, \pm e_2\}} (\omega_h\omega_{h+j} - \bar{\omega}_h\bar{\omega}_{h+j}) \\ &= \beta \sum_{h\in\Lambda_n} \sum_{\substack{h'\in G \\ \{h,h'\}\in C}} (\underbrace{\omega_h\omega_{h'}}_{-1} - \underbrace{\bar{\omega}_h\bar{\omega}_{h'}}_{+1}) \\ &\leq -2\beta|C| . \end{aligned}$$

Now fix $g \in \Lambda_n$ and a simply connected closed curve C such that $g \in I(C) \subseteq \Lambda_n$.

As τ_C is a bijection of E_n^+, it follows that

$$\begin{aligned}
&\sum_{\omega\in E_n^+\cap\Omega_{g,C}} \exp\left(-\beta\psi_n(\omega) - P(-\beta\psi_n, E_n^+)\right)\\
&\leq \sum_{\omega\in E_n^+\cap\Omega_{g,C}} \exp\left(-2\beta|C| - \beta\psi_n(\tau_C\omega) - P(-\beta\psi_n, E_n^+)\right)\\
&\leq \exp\left(-2\beta|C|\right)\cdot\sum_{\omega\in E_n^+} \exp(-\beta\psi_n(\omega))\cdot\exp\left(-P(-\beta\psi_n, E_n^+)\right)\\
&= \exp\left(-2\beta|C|\right). \qquad (5.19)
\end{aligned}$$

Hence

$$\begin{aligned}
&\tilde{\mu}_n\{\omega : \omega_0 = -1\}\\
&= \frac{1}{\lambda_n}\sum_{g\in\Lambda_n} \tilde{\pi}_n\{\omega : \omega_g = -1\}\\
&= \frac{1}{\lambda_n}\sum_{g\in\Lambda_n}\sum_{\omega\in E_n^+,\omega_g=-1} \exp\left(-\beta\psi_n(\omega) - P(-\beta\psi_n, E_n^+)\right)\\
&= \frac{1}{\lambda_n}\sum_{g\in\Lambda_n}\sum_{C\in\mathcal{C}(g)}\sum_{\omega\in E_n^+\cap\Omega_{g,C}} \exp\left(-\beta\psi_n(\omega) - P(-\beta\psi_n, E_n^+)\right)\\
&\leq \frac{1}{\lambda_n}\sum_{g\in\Lambda_n}\sum_{l=4}^{\infty}\sum_{C\in\mathcal{C}(g),|C|=l} \exp\left(-2\beta|C|\right) \quad \text{by (5.19)}\\
&\leq \frac{1}{\lambda_n}\sum_{g\in\Lambda_n}\sum_{l=4}^{\infty} l\cdot 3^{l-1}e^{-2\beta l}\\
&= \sum_{l=4}^{\infty} l\cdot 3^{l-1}e^{-2\beta l}\\
&\to 0 \quad \text{as } \beta\to\infty.
\end{aligned}$$

For the last inequality we had to estimate the number of simply connected closed curves C of length l that contain a given g in their interior. Evidently such a curve contains at least one of the horizontal segments "above g" at distance at most $\frac{l}{2}$ from g, the number of which, by a rough estimate, is bounded by l. Once such a segment is fixed, the rest of the curve is determined by $l-1$ choices among at most three directions.

Hence there is some $\beta_d(\epsilon)$ such that for $\beta > \beta_d(\epsilon)$ and arbitrary n we have $\tilde{\mu}_n\{\omega : \omega_0 = -1\} < \epsilon$. As the measure μ^+ is a weak limit of the $\tilde{\mu}_n$, this yields the desired estimate (5.18). □

5.6 Large deviations for Gibbs measures

As in Sections 5.1–5.3 let ψ be a regular local energy function on the configuration space $\Omega = \Sigma^G$, $G = \mathbb{Z}^d$. $\mathcal{T} = (T^g : g \in G)$ again denotes the shift action.

For $\omega \in \Omega$ and $n > 0$ let

$$\varepsilon_{n,\omega} := \frac{1}{\lambda_n} \sum_{g \in \Lambda_n} \delta_{T^g \omega}$$

be the *empirical distribution* of the orbit $(T^g\omega : g \in \Lambda_n)$. By Birkhoff's ergodic theorem, $\lim_{n\to\infty} \varepsilon_{n,\omega}(f) = \mu(f)$ μ-a.s. for each $f \in C(\Omega)$ and each ergodic $\mu \in \mathcal{M}(\mathcal{T})$. In other words: if $\mu \in \mathcal{M}(\mathcal{T})$ is ergodic, then $\lim_{n\to\infty} \varepsilon_{n,\omega} = \mu$ in the weak topology on $\mathcal{M}$ for μ-a.e. ω. Nevertheless, for finite n, a measure $\varepsilon_{n,\omega}$ may still be quite far from μ. The elementary calculations from Example 1.2.1 suggest that if $\mu \in GS(\psi, \mathcal{T})$ is a Gibbs measure and U a neighbourhood of μ in $\mathcal{M}$, then the probability $\mu\{\omega : \varepsilon_{n,\omega} \notin U\}$ decreases exponentially in n at a rate that depends on the "size" of U. The aim of this section is to make this idea precise. The proofs of our two main propositions are very close to the proof of [62, Theorem 1]. As usual we write $f_n = \sum_{g\in\Lambda_n} f \circ T^g$ for each $f \in C(\Omega)$. With this notation $\varepsilon_{n,\omega}(f) = \frac{1}{\lambda_n} f_n(\omega)$.

Similarly as in Example 1.2.1 we measure the "distance" of an arbitrary $\nu \in \mathcal{M}(\mathcal{T})$ from the set $GS(\psi, \mathcal{T})$ by the quantity

$$d_\psi(\nu) := p(\psi) - (h(\nu) + \nu(\psi)) .$$

Observe that in this way we cannot measure the distance of $\varepsilon_{n,\omega}$ from $GS(\psi, \mathcal{T})$, because $\varepsilon_{n,\omega}$ is in general not invariant. But even if ω is a Λ_n-periodic configuration so that $\varepsilon_{n,\omega}$ is invariant, this would not make sense, because $h(\varepsilon_{n,\omega}) = 0$ and therefore $d_\psi(\varepsilon_{n,\omega}) = p(\psi) - \varepsilon_{n,\omega}(\psi) \to p(\psi) - \mu(\psi) = h(\mu) > 0$ as $n \to \infty$ for μ-a.e. ω if $h(\mu) > 0$. Therefore our actual statements use a modification of this idea.

5.6.1 Theorem (Large deviations for Gibbs measures)
Let ψ be a regular local energy function on $\Omega = \Sigma^G$, $G = \mathbb{Z}^d$, and denote by $\mathcal{T}$ the shift action on Ω. Fix $\mu \in GS(\psi, \mathcal{T})$. For $r > 0$ let $\mathcal{K}_{\psi,r} := \{\nu \in \mathcal{M}(\mathcal{T}) : d_\psi(\nu) \le r\}$. Then $\mathcal{K}_{\psi,r}$ is closed and convex and:

a) *For each open neighbourhood $\mathcal{V}$ of $\mathcal{K}_{\psi,r}$ in $\mathcal{M}$*

$$\limsup_{n\to\infty} \frac{1}{\lambda_n} \log \mu\{\omega : \varepsilon_{n,\omega} \in \mathcal{M} \setminus \mathcal{V}\} \le -r .$$

b) *Suppose that $\mathcal{M}(\mathcal{T}) \setminus \mathcal{K}_{\psi,r} \neq \emptyset$. For each $r' > r$ there is an open neighbourhood $\mathcal{V}$ of $\mathcal{K}_{\psi,r}$ in $\mathcal{M}$ such that*

$$\liminf_{n\to\infty} \frac{1}{\lambda_n} \log \mu\{\omega : \varepsilon_{n,\omega} \in \mathcal{M} \setminus \mathcal{V}\} \ge -r' .$$

5.6.2 Remark For $m > 0$ denote by $\Omega_m = \{+1, -1\}^{\Lambda_m}$ the set of $(+,-)$-configurations in the Ising model that can be observed in the finite "window" Λ_m. Denote by ν_m the canonical projection of a measure $\nu \in \mathcal{M}(T)$ to Ω_m. Then ν_m is a probability distribution on the finite space Ω_m. Now the first part of the theorem can be interpreted as follows: Let μ^+ be the equilibrium state for the Ising model favouring the "+". Consider $\omega \in \Omega$. We screen ω by sliding the window Λ_m over the configuration within the bounds of Λ_n $(n > m)$ and record the empirical distribution of local configurations seen inside this window. The μ^+-probability that this empirical distribution is significantly different from all ν_m where ν are equilibrium states for the Ising model decreases as e^{-cn^2}. The constant c depends on how we specify "significantly". On the other hand, the probability that this empirical distribution is close to μ_m^- (which is significantly different from μ_m^+) is, in view of (5.8), at least of the order $e^{-C_\psi n}$. This means that although on a large grid Λ_n configurations with dominating "+" in a "window" Λ_m are most likely under $\mu = \mu^+$, configurations with dominating "−" that are distributed according to μ^- are still much more likely than configurations with other structures like those with approximately equal numbers of "+" and "−". This offers an explanation for the occurrence of large "−" islands in an ocean of "+" in μ^+-typical configurations. And of course, these "−" islands contain again relatively large "+" ponds that contain not too small "−" islands and so on. This situation is illustrated in Figure 5.1 on p. 111. ◇

The proof of the theorem rests on upper and lower estimates of the type one usually finds in large deviations theory.

5.6.3 Proposition (Upper large deviations estimate) *Let $\mu \in GS(\psi, T)$, $f \in C(\Omega)$ and $\gamma \in \mathbb{R}$. Then*

$$\limsup_{n\to\infty} \frac{1}{\lambda_n} \log \mu\{\omega : \varepsilon_{n,\omega}(f) \geq \gamma\} \leq -\inf\{d_\psi(\nu) : \nu \in \mathcal{M}(T),\ \nu(f) \geq \gamma\}.$$

(The infimum on the r.h.s. of this inequality is just the "distance" of $\{\nu : \nu(f) \geq \gamma\}$ from $GS(\psi, T)$.)

Proof: Let $B_n := \{\omega : \varepsilon_{n,\omega}(f) \geq \gamma\}$. We will produce a measure $\nu \in \mathcal{M}(T)$ with $\nu(f) \geq \gamma$ and such that

$$\limsup_{n\to\infty} \frac{1}{\lambda_n} \log \mu(B_n) \leq -d_\psi(\nu).$$

To this end let $\mathcal{Z}_n := \{A \in \alpha^{\Lambda_n} : A \cap B_n \neq \emptyset\}$. Here $\alpha = \{[\sigma]_0 : \sigma \in \Sigma\}$ (see Example 3.2.20). For each $A \in \mathcal{Z}_n$ fix a point $\omega_A \in A \cap B_n$. Then $E_n := \{\omega_A : A \in \mathcal{Z}_n\}$ is a maximal $(n, 1)$-separated subset of B_n, see Example 4.4.2. Because of Lemma 4.4.8, which prepared the construction of equilibrium states, there is a measure $\nu \in \mathcal{M}(\mathcal{T})$ with

$$h(\nu) + \nu(\psi) \geq \limsup_{n\to\infty} \frac{1}{\lambda_n} P(\psi_n, E_n) \,. \tag{5.20}$$

By construction it is the weak limit of a subsequence of the measures

$$\nu_n = \frac{1}{\lambda_n} \sum_{g\in\Lambda_n} \sum_{\omega\in E_n} e^{\psi_n(\omega)-P(\psi_n,E_n)} \cdot \delta_{T^g\omega} = \sum_{\omega\in E_n} e^{\psi_n(\omega)-P(\psi_n,E_n)} \cdot \varepsilon_{n,\omega} \,.$$

As $E_n \subseteq B_n$, each ν_n is a convex combination of measures $\varepsilon_{n,\omega}$ with $\varepsilon_{n,\omega}(f) \geq \gamma$. Hence $\nu(f) \geq \gamma$.

Recall that $e^{P(\psi_n,E_n)} = \sum_{\omega\in E_n} e^{\psi_n(\omega)}$. Hence, by (5.8) of Theorem 5.2.4,

$$\mu(B_n) \leq \sum_{A\in\mathcal{Z}_n} \mu(A) \leq \sum_{\omega\in E_n} e^{\psi_n(\omega)-P_{n,1}(\psi)+C_\psi n^{d-1}} = e^{P(\psi_n,E_n)-P_{n,1}(\psi)+C_\psi n^{d-1}}$$

and therefore

$$\limsup_{n\to\infty} \frac{1}{\lambda_n} \log \mu(B_n) \leq h(\nu) + \nu(\psi) - p(\psi) = -d_\psi(\nu)$$

where we used (5.20) and Corollary 5.3.3. □

5.6.4 Proposition (Lower large deviations estimate) *Let $\mu \in GS(\psi, \mathcal{T})$, $f \in C(\Omega)$ and $\gamma \in \mathbb{R}$. Then*

$$\liminf_{n\to\infty} \frac{1}{\lambda_n} \log \mu\{\omega : \varepsilon_{n,\omega}(f) > \gamma\} \geq -\inf\{d_\psi(\nu) : \nu \in \mathcal{M}(\mathcal{T}),\, \nu(f) > \gamma\} \,.$$

Proof: The proof consists of three parts. In the first one, which is purely ergodic theoretic, the inequality is proved for ergodic ν. The second step is a combinatorial argument that extends the result to certain ν which are finite rational convex combinations of ergodic measures. The final step is to approximate arbitrary ν by such convex combinations. For that step we refer to Exercise 4.3.10, because the estimate for arbitrary ν is not needed in the proof of Theorem 5.6.1.

Step 1: Let $B_n := \{\omega : \epsilon_{n,\omega}(f) > \gamma\}$. Suppose that $\nu \in \mathcal{M}(\mathcal{T})$ is ergodic. Let $\epsilon > 0$ and set

$$\mathcal{Z}_n := \{A \in \alpha^{\Lambda_n} : \exists \omega \in A \text{ with } \frac{1}{\lambda_n} f_n(\omega) > \nu(f) - \epsilon,\, \frac{1}{\lambda_n}\psi_n(\omega) > \nu(\psi) - \epsilon\} \,. \tag{5.21}$$

Because of Lemma 5.1.4 there is $n_0 > 0$ (independent of ν) such that

$$\frac{1}{\lambda_n} f_n(\omega) > \nu(f) - 2\epsilon \quad \text{and} \quad \frac{1}{\lambda_n}\psi_n(\omega) > \nu(\psi) - 2\epsilon \tag{5.22}$$

for all $n \geq n_0$, $A \in \mathcal{Z}_n$ and $\omega \in A$.

Let $q_n := \nu(\bigcup \mathcal{Z}_n)$. Then $\lim_{n\to\infty} q_n = 1$ by Birkhoff's ergodic theorem. Denote by ν'_n, ν''_n the normalized restrictions of ν to $\bigcup \mathcal{Z}_n$ and $\Omega \setminus \bigcup \mathcal{Z}_n$ respectively. Then

$$\begin{aligned} H_\nu(\alpha^{\Lambda_n}) &\leq q_n\, H_{\nu'_n}(\alpha^{\Lambda_n}) + (1-q_n)\, H_{\nu''_n}(\alpha^{\Lambda_n}) + \log 2 \\ &\leq q_n \cdot \log \operatorname{card}(\mathcal{Z}_n) + (1-q_n)\cdot \log \operatorname{card}(\alpha^{\Lambda_n}) + \log 2 \end{aligned}$$

by Theorem 3.3.2. Hence

$$\begin{aligned} h(\nu) &\leq \liminf_{n\to\infty} \frac{q_n}{\lambda_n} \log \operatorname{card}(\mathcal{Z}_n) + \lim_{n\to\infty} (1-q_n) \log \operatorname{card}(\alpha) \\ &= \liminf_{n\to\infty} \frac{1}{\lambda_n} \log \operatorname{card}(\mathcal{Z}_n)\,. \end{aligned} \tag{5.23}$$

Next, by (5.8) of Theorem 5.2.4,

$$\begin{aligned} \log \mu(A) &\geq \psi_n(\omega) - P_{n,1}(\psi) - C_\psi n^{d-1} \\ &\geq \lambda_n\,(\nu(\psi) - 2\epsilon) - P_{n,1}(\psi) - C_\psi n^{d-1} \end{aligned} \tag{5.24}$$

for each $A \in \mathcal{Z}_n$, $n \geq n_0$, and $\omega \in A$, where we used (5.22) for the second inequality. Together with (5.23) and Corollary 5.3.3 this gives

$$\liminf_{n\to\infty} \frac{1}{\lambda_n} \log \mu\left(\bigcup \mathcal{Z}_n\right) \geq h(\nu) + \nu(\psi) - 2\epsilon - p(\psi) = -d_\psi(\nu) - 2\epsilon\,.$$

Now it is easy to finish the proof of the proposition for ergodic ν: Suppose that $\nu(f) > \gamma$ and $2\epsilon < \nu(f) - \gamma$. Then $B_n \supseteq \bigcup \mathcal{Z}_n$ for sufficiently large n by (5.22) so that $\liminf_{n\to\infty} \frac{1}{\lambda_n} \log \mu(B_n) \geq -d_\psi(\nu) - 2\epsilon$ for each $\epsilon > 0$.

Step 2: Suppose now that $\nu = \sum_{i=1}^N a_i \nu_i$ where the ν_i are ergodic and $a_i = \frac{k_i}{\lambda_s}$ for some $k_1, \ldots, k_N, s \in \mathbb{N}$. Set $\ell = \ell(n) := 2sn - s - n + 1$. In analogy to the choice of $\mathcal{Z}_n$ in the first step, we are going to construct a family $\mathcal{Z}_{s,n}$ of cylinder sets $A \in \alpha^{\Lambda_\ell}$ containing configurations ω which are "typical" for ν. To this end we decompose Λ_ℓ into λ_s disjoint copies of Λ_n and require our cylinders to contain just those configurations which are typical for ν_i in the sense of (5.22) on exactly k_i of these copies.

Fix a map $j : \Lambda_s \to \{1, \ldots, N\}$ such that $\operatorname{card} j^{-1}\{i\} = k_i$ for all $i = 1, \ldots, N$ and let

$$\mathcal{Z}_{s,n} := \left\{ \bigcap_{g\in\Lambda_s} T^{-(2n-1)g} A_g : A_g \in \mathcal{Z}_n(\nu_{j(g)}) \text{ for all } g \in \Lambda_s \right\}.$$

Here $\mathcal{Z}_n(\nu)$ denotes the set $\mathcal{Z}_n \subseteq \alpha^{\Lambda_n}$ constructed for an ergodic measure ν in the first step. Then $\mathcal{Z}_{s,n} \subseteq \alpha^{\Lambda_\ell}$, $\lambda_\ell = (4sn - 2s - 2n + 1)^d = \lambda_s \lambda_n$, and

$$\operatorname{card} \mathcal{Z}_{s,n} = \prod_{g\in\Lambda_s} \operatorname{card} \mathcal{Z}_n(\nu_{j(g)}) = \prod_{i=1}^N (\operatorname{card} \mathcal{Z}_n(\nu_i))^{k_i}\,.$$

In view of (5.23) this implies

$$\begin{aligned}\liminf_{n\to\infty}\frac{1}{\lambda_s\lambda_n}\log\operatorname{card}\left(\mathcal{Z}_{s,n}\right) &\geq \sum_{i=1}^{N}\frac{k_i}{\lambda_s}\liminf_{n\to\infty}\frac{1}{\lambda_n}\log\operatorname{card}\mathcal{Z}_n(\nu_i)\\ &\geq \sum_{i=1}^{N}a_i h(\nu_i)=h(\nu) \qquad (5.25)\end{aligned}$$

where the last identity follows from Theorem 3.3.2.

Next consider $\psi_\ell(\omega)$ for $A\in\mathcal{Z}_{s,n}$ and $\omega\in A$. If $n\geq n_0$, then

$$\begin{aligned}\psi_\ell(\omega) &= \sum_{g\in\Lambda_s}\psi_n(T^{(2n-1)g}\omega)\geq\sum_{g\in\Lambda_s}\lambda_n\cdot(\nu_{j(g)}(\psi)-2\epsilon)\\ &= \lambda_n\sum_{i=1}^{N}k_i\cdot(\nu_i(\psi)-2\epsilon) \qquad (5.26)\end{aligned}$$

as $T^{(2n-1)g}\omega\in A_g\in\mathcal{Z}_n(\nu_{j(g)})$ for all $g\in\Lambda_s$. Therefore, as in (5.24), we have for each $A\in\mathcal{Z}_{s,n}$, $n\geq n_0$, and $\omega\in A$

$$\begin{aligned}\log\mu(A) &\geq \psi_\ell(\omega)-P_{\ell,1}(\psi)-C_\psi\ell^{d-1}\\ &\geq \lambda_n\sum_{i=1}^{N}k_i\cdot(\nu_i(\psi)-2\epsilon)-P_{\ell,1}(\psi)-C_\psi\ell^{d-1}\\ &= \lambda_n\lambda_s(\nu(\psi)-2\epsilon)-P_{\ell,1}(\psi)-C_\psi\ell^{d-1}\,.\end{aligned}$$

Combining this with the previous estimate (5.25) we get

$$\liminf_{n\to\infty}\frac{1}{\lambda_{\ell(n)}}\log\mu\left(\bigcup\mathcal{Z}_{s,n}\right)\geq h(\nu)+\nu(\psi)-p(\psi)-2\epsilon=-d_\psi(\nu)-2\epsilon$$

because of $\lambda_s\lambda_n=\lambda_\ell$ and Corollary 5.3.3.

Now the proof is finished as for ergodic ν: Suppose that $\nu(f)>\gamma$ and $3\epsilon<\nu(f)-\gamma$. Exactly as in (5.26) one proves that for $n\geq n_0$ and $\omega\in A\in\mathcal{Z}_{s,n}$

$$f_{\ell(n)}(\omega)\geq\lambda_n\sum_{i=1}^{N}k_i\cdot(\nu_i(f)-2\epsilon)=\lambda_{\ell(n)}\cdot(\nu(f)-2\epsilon)>\lambda_{\ell(n)}\cdot(\gamma+\epsilon)\,.$$

As, by the mean value theorem of differentiation,

$$\begin{aligned}\frac{\lambda_{\ell(n+1)}-\lambda_{\ell(n)}}{\lambda_{\ell(n)}} &\leq \frac{(4s-2)\,d\,(4s(n+1)-2s-2(n+1)+1)^{d-1}}{\lambda_{\ell(n)}}\\ &= \frac{(4s-2)\,d}{4sn-2s-2n+1}\left(1+\frac{4s-2}{4sn-2s-2n+1}\right)^{d-1},\end{aligned}$$

there is $n_1 \geq 0$ such that for $n \geq n_1$

$$\frac{\lambda_{\ell(n+1)} - \lambda_{\ell(n)}}{\lambda_{\ell(n)}} < \frac{\epsilon}{\|f\|_\infty} .$$

Therefore, if $n \geq n_1$ and $\ell(n) \leq k < \ell(n+1)$, then

$$f_k(\omega) \geq f_{\ell(n)}(\omega) - (\lambda_{\ell(n+1)} - \lambda_{\ell(n)}) \cdot \|f\|_\infty > \lambda_{\ell(n)} \cdot \gamma$$

and

$$f_k(\omega) \geq f_{\ell(n+1)}(\omega) - (\lambda_{\ell(n+1)} - \lambda_{\ell(n)}) \cdot \|f\|_\infty > \lambda_{\ell(n+1)} \cdot \gamma .$$

It follows that $f_k(\omega) > \lambda_k \cdot \gamma$, so that each $A \in \mathcal{Z}_{s,n}$ is contained in B_k for such k and sufficiently large n. Therefore

$$\liminf_{k\to\infty} \frac{1}{\lambda_k} \log \mu(B_k) \geq \liminf_{n\to\infty} \frac{1}{\lambda_{\ell(n)}} \log \mu\left(\bigcup \mathcal{Z}_{s,n}\right) \geq -d_\psi(\nu) - 2\epsilon$$

for each $\epsilon > 0$. □

Proof of Theorem 5.6.1:
a) Let $\mathcal{V}$ be an open neighbourhood of $\mathcal{K}_{\psi,r} = \{\nu \in \mathcal{M}(\mathcal{T}) : d_\psi(\nu) \leq r\}$. As $\mathcal{M}(\mathcal{T})$ is a closed convex subset of $\mathcal{M}$ and as $d_\psi : \mathcal{M}(\mathcal{T}) \to \mathbb{R}$ is lower semicontinuous and convex affine (see Theorem 3.3.2), $\mathcal{K}_{\psi,r}$ is compact and convex.

Let $\nu \in \mathcal{M} \setminus \mathcal{V}$. As in the proof of Theorem 4.2.9 one shows that ν can be separated from $\mathcal{K}_{\psi,r}$ by some hyperplane, i.e., there exists $f_\nu \in C(\Omega)$ such that $\nu(f_\nu) > 0$ and $\tilde{\nu}(f_\nu) < 0$ for each $\tilde{\nu} \in \mathcal{K}_{\psi,r}$. Let $\mathcal{H}_\nu := \{\tilde{\nu} \in \mathcal{M} : \tilde{\nu}(f_\nu) > 0\}$ be the open half space containing ν. Then $\bar{\mathcal{H}}_\nu \subset \mathcal{M} \setminus \mathcal{K}_{\psi,r}$. As $\mathcal{M} \setminus \mathcal{V}$ is compact, there are $\nu_1, \ldots, \nu_N \in \mathcal{M} \setminus \mathcal{V}$ such that $\mathcal{M} \setminus \mathcal{V} \subseteq \bigcup_{i=1}^N \mathcal{H}_{\nu_i}$. Therefore

$$\begin{aligned}
&\limsup_{n\to\infty} \frac{1}{\lambda_n} \log \mu\{\omega : \varepsilon_{n,\omega} \in \mathcal{M} \setminus \mathcal{V}\} \\
&\leq \limsup_{n\to\infty} \frac{1}{\lambda_n} \log \sum_{i=1}^N \mu\{\omega : \varepsilon_{n,\omega} \in \mathcal{H}_{\nu_i}\} \\
&\leq \max_{i=1,\ldots,N} \limsup_{n\to\infty} \frac{1}{\lambda_n} \log \mu\{\omega : \epsilon_{n,\omega}(f_{\nu_i}) > 0\} \\
&\leq \max_{i=1,\ldots,N} \left(-\inf\{d_\psi(\tilde{\nu}) : \tilde{\nu} \in \bar{\mathcal{H}}_{\nu_i} \cap \mathcal{M}(\mathcal{T})\}\right) \quad \text{by Proposition 5.6.3} \\
&\leq -\inf\{d_\psi(\tilde{\nu}) : \tilde{\nu} \in \mathcal{M}(\mathcal{T}) \setminus \mathcal{K}_{\psi,r}\} \\
&\leq -r .
\end{aligned}$$

b) Let $r' > r$. As $\mathcal{K}_{\psi,r} \neq \mathcal{M}(\mathcal{T})$ and as $\mathcal{K}_{\psi,r}$ is convex, there exists an ergodic $\nu_0 \in \mathcal{M}(\mathcal{T}) \setminus \mathcal{K}_{\psi,r}$ (observe Theorem 2.3.3), so $d_\psi(\nu_0) > r$. On the other hand, there is some ergodic $\mu \in GS(\psi, \mathcal{T}) = ES(\psi, \mathcal{T})$, so $d_\psi(\mu) = 0$. Let

$$\nu_t := t\mu + (1-t)\nu_0 \quad (0 \leq t \leq 1) .$$

Then $d_\psi(\nu_t) = (1-t)d_\psi(\nu_0)$ as $\nu \mapsto d_\psi(\nu)$ is convex affine (see Theorem 3.3.2). Fix $a \in (0,1)$ such that $r < d_\psi(\nu_a) < r'$. Then $\nu_a \notin \mathcal{K}_{\psi,r}$, and as $\mathcal{K}_{\psi,r}$ is compact and convex, ν_a can be separated from $\mathcal{K}_{\psi,r}$ by some hyperplane as in the first part of the proof. Denote by $\mathcal{H}_{\nu_a}$ the half space containing ν_a and by $\mathcal{V}$ the interior of $\mathcal{M} \setminus \mathcal{H}_{\nu_a}$. Then $\mathcal{V}$ is a neighbourhood of $\mathcal{K}_{\psi,r}$ and

$$
\begin{aligned}
&\liminf_{n\to\infty} \frac{1}{\lambda_n} \log \mu\{\omega : \varepsilon_{n,\omega} \in \mathcal{M} \setminus \mathcal{V}\} \\
&\quad \geq \liminf_{n\to\infty} \frac{1}{\lambda_n} \log \mu\{\omega : \varepsilon_{n,\omega} \in \mathcal{H}_{\nu_a}\} \\
&\quad \geq -\inf\{d_\psi(\tilde{\nu}) : \tilde{\nu} \in \mathcal{H}_{\nu_a} \cap \mathcal{M}(T)\} \quad \text{by Proposition 5.6.4} \\
&\quad \geq -d_\psi(\nu_a) \\
&\quad > -r' .
\end{aligned}
$$

Observe that we can choose a rational and of the special form considered in the second step of the proof of Proposition 5.6.4 so that the version of that proposition which we actually proved is sufficient for the present proof. □

Chapter 6

Equilibrium states and derivatives

The natural model for a dynamical system in discrete time is a map T acting on a space X, see Section 1.4. Quite often the underlying space X possesses a natural volume measure m. If X is a subset of a Euclidean space, m may be Lebesgue measure, and, more generally, if X is part of a Riemannian manifold, m may be the natural Riemannian volume. If the action respects this volume, i.e., if m-null sets are mapped to m-null sets, a "relevant" invariant measure should reflect the asymptotic behaviour of a set of points of positive volume in the sense that $\lim_{n\to\infty}\frac{1}{n}\sum_{k=0}^{n-1}\phi(T^kx) = \int \phi\, d\mu$ for each $\phi \in C(X)$ and for a set of points x of positive m-measure. Such measures are called *Sinai–Bowen–Ruelle measures* (SBR measures). Ergodic invariant measures which are absolutely continuous with respect to m belong to this class, but also the point mass in a stable fixed point of T is a SBR measure. It is a most interesting fact that in many cases SBR measures can be characterized by a variational principle, and for particularly regular locally expanding maps the only SBR measures are the absolutely continuous ones (w.r.t. m). The first three sections of this chapter, which are centred around these ideas, are influenced by various publications, among them [36] and [57]. We do not treat applications to hyperbolic maps in this book, i.e., to maps having expanding and contracting directions. Readers interested in this topic are referred to [6], [37], [47] and, for a broader view on the theory of dynamical systems, to [31]. Iterated function systems, however, which bear some resemblance to hyperbolic maps, are studied in the last section, where the Hausdorff dimension of their associated fractal sets is related to the variational formalism.

In this whole chapter, except for the section on iterated function systems, we assume that $G = \mathbb{Z}_+$, i.e., that the action $\mathcal{T}$ is generated by a (typically noninvertible) map T. Instead of $ES(\psi, \mathcal{T})$ we write $ES(\psi, T)$.

6.1 Sinai–Bowen–Ruelle measures

In this section we characterize SBR measures μ in terms of the quantity $h(\mu)+\mu(\psi)$ where $-\psi$ is the logarithm of the volume derivative of T w.r.t. m. As some of our considerations are of purely measure theoretic nature, we work in the following general setting:

▷ $(X, \mathcal{B}, m)$ is a probability space,

▷ α is a finite $\mathcal{B}$-measurable partition of X,

▷ $T : X \to X$ is a $\mathcal{B}$-measurable transformation such that all $T_{|A} : A \to T(A)$ $(A \in \alpha)$ are bijective, bimeasurable and nonsingular w.r.t. m, i.e., $m(TB) = 0$ if and only if $m(B) = 0$ for every $B \in \mathcal{B}$, $B \subseteq A$.

We call such a quintuple $(X, \mathcal{B}, m, T, \alpha)$ a *fibred system*. (This is a slight variation of the usage of this term in [55] where the partition α can be countable. In Section 6.3 we also allow countable partitions.) Typically the measure m is not T-invariant. Instead we use that T is "conformal" with respect to m in the sense made precise in the following definition:

6.1.1 Definition *Let $(X, \mathcal{B}, m, T, \alpha)$ be a fibred system and let $\psi : X \to [-\infty, \infty]$ be $\mathcal{B}$-measurable. The system is $e^{-\psi}$-conformal, if*

$$m(TB) = \int_B e^{-\psi}\, dm \text{ for all } B \in \mathcal{B},\ B \subseteq A \in \alpha. \tag{6.1}$$

$(X, \mathcal{B}, m, T, \alpha)$ is a continuous $e^{-\psi}$-conformal fibred system, if X is compact metric with Borel σ-algebra $\mathcal{B}$, if $T : X \to X$ is continuous and if $\psi : X \to [-\infty, \infty)$ is upper semicontinuous.

The classical situation is that $X \subseteq \mathbb{R}^d$ has positive finite Lebesgue measure, $\mathcal{B}$ is the Borel σ-algebra of X, m is Lebesgue measure on X, $T : X \to X$ is piecewise differentiable (on the "pieces" $A \in \alpha$) and $\psi = -\log|\det(DT)|$. More generally, if $(X, \mathcal{B}, m, T, \alpha)$ is $e^{-\psi}$-conformal, then $e^{-\psi}$ is a kind of volume derivative with respect to the measure m.

6.1.2 Remarks

a) $\int_A f \cdot e^{-\psi}\, dm = \int_{TA} f \circ T_{|A}^{-1}\, dm$ for $A \in \alpha$ and measurable $f : X \to [0, \infty]$.

b) $m\{x : \psi(x) = +\infty\} = 0$, because $m(T(B \cap A)) = \int_{B\cap A} e^{-\psi}\, dm = 0$ for $B = \{x : \psi(x) = +\infty\}$ and $A \in \alpha$, so that also $m(B) = 0$ since T is noninvertible.

c) $m\{x : \psi(x) = -\infty\} = 0$, because $\int_{B\cap A} e^{-\psi}\, dm = m(T(B\cap A)) \le m(X) = 1$ for $B = \{x : \psi(x) = -\infty\}$ and $A \in \alpha$.

d) There is a chain rule for $e^{-\psi}$-conformal systems, namely

$$m(T^n B) = \int_B e^{-\psi_n}\, dm$$

for measurable $B \subseteq A \in \alpha_0^n$. ◇

6.1.3 Exercise Show that for each fibred system $(X, \mathcal{B}, m, T, \alpha)$ there is some $\mathcal{B}$-measurable $\psi : X \to [-\infty, \infty]$ such that the system is $e^{-\psi}$-conformal. ◇

6.1.4 Example Let X be an interval, α a m-partition of X into subintervals, and suppose that $T_{|A}$ is strictly monotonic and differentiable for each $A \in \alpha$. Then $(X, \mathcal{B}, m, T, \alpha)$ is $|T'|$-conformal. We call such maps *piecewise monotonic*. Here is a list of examples:

1. *Circle maps*
   Let $r \in \{2, 3, 4, \ldots\}$ and let $\hat{T} : [0,1] \to [0,r]$ with $\hat{T}(0) = 0$ and $\hat{T}(1) = r$ be strictly monotone, continuous, and differentiable except at at most countably many points. Define $T : [0,1) \to [0,1)$ by $T(x) = \hat{T}(x) \bmod 1$. Identifying the endpoints of the interval we can interpret T as a continuous *r-fold covering circle map*. Observe that T is a local homeomorphism. Particular examples are
   a) $T_\theta x = \frac{\theta x}{1+(\theta-2)x}$ if $0 \le x < \frac{1}{2}$, $T_\theta x = 1 - T_\theta(1-x)$ if $\frac{1}{2} < x \le 1$ with $0 < \theta < \infty$,
   b) $Tx = 1 - \sqrt{1-2x}$ if $0 \le x < \frac{1}{2}$, $Tx = 1 - T(1-x)$ if $\frac{1}{2} < x \le 1$.
   In both of these cases T has two monotone branches mapping onto $[0,1]$. Observe that $T_\theta'(0) = T_\theta'(1) = \theta$.
2. *Symmetric tent maps*
   $T_s : [0,1] \to [0,1]$, $T_s x = s \cdot (\frac{1}{2} - |x - \frac{1}{2}|)$ with $0 < s \le 2$.
3. *Logistic maps*
   $T_a : [0,1] \to [0,1]$, $T_a x = ax(1-x)$, $2 < a \le 4$.

We note already now that logistic maps with parameters $a \neq 4$ cannot be treated by the methods developed in this chapter. The interested reader is referred to [40]. ◇

In order to keep the discussion as elementary as possible we do not consider specific examples of higher-dimensional maps in this book. Nevertheless, most theorems are formulated in such a way that they can be applied in more complex situations, although it may be much harder to check their assumptions.

6.1.5 Remark How piecewise continuous T and piecewise upper semicontinuous ψ can be fitted into this framework is discussed in Section A.5, see also Remark 1.4.6. Among others, for piecewise continuous interval maps with finitely many monotone branches the following turns out: By adding countably many points to X and changing the topology in a suitable way, T can be made continuous and ψ upper semicontinuous. In most cases there is no periodic point among the countably many new ones so that the enlarged system has exactly the same invariant measures as the original one. Otherwise the only new invariant measures that can arise in this way are equidistributions on periodic orbits, see also Remark 6.1.10. ◇

6.1.6 Definition *Let $(X, \mathcal{B}, m, T, \alpha)$ be a continuous fibred system. For $x \in X$ denote by $V_T(x)$ the set of all accumulation points of the sequence of measures $(\frac{1}{n}\sum_{k=0}^{n-1}\delta_{T^k x})_{n>0}$.*

1. *A measure $\mu \in \mathcal{M}$ is observable, if*

$$m\{x \in X : \mu \in V_T(x)\} > 0 .$$

2. *A measure $\mu \in \mathcal{M}$ is a Sinai–Bowen–Ruelle measure (SBR measure) if*

$$m\{x \in X : V_T(x) = \{\mu\}\} > 0 .$$

Recall from the proof of Theorem 1.4.3 that $V_T(x)$ is a nonempty subset of $\mathcal{M}(T)$. In particular, observable measures are T-invariant.

6.1.7 Remarks

a) Observable measures need not exist. The simplest example is the identity map on $[0, 1]$ equipped with Lebesgue measure. In this case $V_T(x) = \{\delta_x\}$ for each x.

b) More generally, the following consideration may show that a map T has no observable measures: Let $\mu = \int \mu_x \, d\mu(x)$ be the ergodic decomposition of $\mu \in \mathcal{M}(T)$. Birkhoff's ergodic theorem applied to a countable dense set D of functions $f \in C(X)$ shows that there is a set $X_0 \subseteq X$ of full μ-measure such that

$$\lim_{n\to\infty}\frac{1}{n}\sum_{k=0}^{n-1}\delta_{T^k x}(f) = E_\mu[f \mid \mathcal{I}_\mu(T)](x) = \mu_x(f) \quad \text{for all } f \in D \text{ and all } x \in X_0.$$

Hence $\lim_{n\to\infty}\frac{1}{n}\sum_{k=0}^{n-1}\delta_{T^k x} = \mu_x$ for all $x \in X_0$ which can be rephrased as $V_T(x) = \{\mu_x\}$ for μ-a.e. x. Thus, if $\mu \approx m$ and if $\mathcal{I}_\mu(T)$ contains no atoms, then T has no observable measure. (For nontrivial examples see Exercise 2.3.4.) ◇

Given an $e^{-\psi}$-conformal fibred system $(X, \mathcal{B}, m, T, \alpha)$ we write

$$F_\alpha : \mathcal{M}(T) \to [-\infty, \infty), \quad \nu \mapsto h(\nu, \alpha, T) + \nu(\psi) ,$$
$$F : \mathcal{M}(T) \to [-\infty, \infty), \quad \nu \mapsto h(\nu) + \nu(\psi) .$$

With this notation $p(\psi) = \sup_\mu F(\mu)$ if $(X, \mathcal{B}, m, T, \alpha)$ is continuous. $F(\mu)$ is sometimes called the *free energy* of μ. Observe also that $\nu(\psi)$ is the *Lyapunov exponent* of ν, see Section 1.4.

6.1.8 Theorem (Free energy of observable or absolutely continuous measures) *Let $(X, \mathcal{B}, m, T, \alpha)$ be a continuous $e^{-\psi}$-conformal fibred system. Suppose that for all $\mu \in \mathcal{M}(T)$*

$$\lim_{n\to\infty} \frac{1}{n} \left(\sum_{A\in\alpha_0^n} \mu(\bar{A} \setminus A) \,|\log \mu(A)| + \log a_n \right) = 0 \tag{6.2}$$

where $\bar{A}$ is the topological closure of A and $a_n := |\{A \in \alpha_0^n : \mu(\bar{A}) > e^{-1}\}| + 1$. Then

a) *$V_T(x) \subseteq \{\mu \in \mathcal{M}(T) : F_\alpha(\mu) \geq 0\}$ for m-a.e. x.*

b) *$F_\alpha(\mu) \geq 0$ for each observable measure μ.*

c) *Suppose that $\mu \in \mathcal{M}(T)$ is absolutely continuous w.r.t. m and denote by $\mu = \int \mu_x \, d\mu(x)$ the ergodic decomposition of μ. Then $F_\alpha(\mu_x) \geq 0$ for μ-a.e. x. In particular $F(\mu) \geq 0$.*

6.1.9 Remark We note already here that F_α and F are affine convex by Theorem 3.3.2 and upper semicontinuous by Theorem 4.2.4 and Corollary 4.1.7. ◇

6.1.10 Remark In the case of piecewise monotonic interval maps it is not difficult to verify assumption (6.2): Denote by E the set of endpoints of all $A \in \alpha$ and let $\tilde{E} := \bigcup_{k=0}^{\infty} T^{-k}E$. Then $\tilde{E}$ is countable and $\bar{A} \setminus A \subset \tilde{E}$ for each $A \in \alpha_0^n$, because these A are intervals whose endpoints belong to $\tilde{E}$. Therefore, the only way in which $\mu(\bar{A} \setminus A)$ can be positive is that $\mu\{x\} > 0$ for some $x \in \bar{A} \cap \tilde{E}$. As $\mu\{T^{k+l}x\} = \mu(T^{-l}\{T^{k+l}x\}) \geq \mu\{T^k x\} \geq \mu\{x\} > 0$ for all $k, l \in \mathbb{N}$, there are $k \geq 0$ and $l > 0$ such that $T^l(T^k x) = T^k x$. If $k > 0$, then $\mu\{T^k x\} = \mu(T^{-1}\{T^k x\}) \geq \mu\{T^{k-1}x\} + \mu\{T^{k+l-1}x\} \geq \mu\{x\} + \mu\{T^k x\}$, a contradiction to $\mu\{x\} > 0$. Therefore, $T^l x = x$ and, as $x \in \tilde{E}$, the orbit of x intersects E. As E is finite there are at most finitely many such periodic orbits. Hence there is a bound on the number of nonzero summands in (6.2) which is uniform in n. As for each point y there are at most two $A \in \alpha_0^n$ such that $y \in \bar{A} \setminus A$, the sequence $(a_n)_{n\in\mathbb{N}}$ is also uniformly bounded (namely by $2e + 1$). ◇

The theorem is a corollary to the following large deviations estimate: As in Section 5.6 let

$$\varepsilon_{n,x} := \frac{1}{n} \sum_{k=0}^{n-1} \delta_{T^k x}$$

be the *empirical distribution* of the orbit $(T^k x : k = 0, \ldots, n-1)$. For $r \geq 0$ let

$$\mathcal{K}_r := \{\nu \in \mathcal{M}(T) : F_\alpha(\nu) \geq -r\} .$$

Obviously $\mathcal{K}_r$ is compact and convex.

6.1.11 Proposition (Upper large deviations for fibred systems) *Let $\mathcal{V}$ be an open neighbourhood of $\mathcal{K}_r$ in $\mathcal{M}$. Under the assumptions of Theorem 6.1.8*

$$\limsup_{n\to\infty} \frac{1}{n} \log m\{x : \varepsilon_{n,x} \in \mathcal{M} \setminus \mathcal{V}\} \leq -r\,. \tag{6.3}$$

Proof of Theorem 6.1.8:

a) Let $r > 0$. The proposition together with the Borel–Cantelli lemma implies that for m-a.e. x and each open neighbourhood $\mathcal{V}$ of $\mathcal{K}_r$ there is $n_0 > 0$ such that $\varepsilon_{n,x} \in \mathcal{V}$ for all $n \geq n_0$. Hence $V_T(x) \subseteq \mathcal{K}_r$ for m-a.e. x. It follows that $V_T(x) \subseteq \bigcap_{k\in\mathbb{N}} \mathcal{K}_{2^{-k}} = \mathcal{K}_0 = \{\mu \in \mathcal{M}(T) : F_\alpha(\mu) \geq 0\}$ for m-a.e. x.

b) If μ is observable, then $\mu \in V_T(x) \subseteq \mathcal{K}_0$ for a nonempty set of points x.

c) Let $\mu = \int \mu_x \, d\mu(x)$ be the ergodic decomposition of μ. Then $V_T(x) = \{\mu_x\}$ for μ-a.e. x by Remark 6.1.7-b and, as $\mu \ll m$, assertion a) of this theorem implies that $F(\mu_x) \geq F_\alpha(\mu_x) \geq 0$ for μ-a.e. x. Observing Theorem 4.3.7 and Remark 6.1.9 it follows that $F(\mu) = \int F(\mu_x)\, d\mu(x) \geq 0$. □

Proof of the proposition: The proof is similar to the upper large deviations estimate in Section 5.6. The main difference is that the role of the Gibbs state μ in that proof is now played by the measure m, the approximation formula (5.8) for the μ-measure of cylinders $A \in \alpha^{\Lambda_n}$ being replaced by the conformality property (6.1).

Let $\mathcal{V}$ be an open neighbourhood of $\mathcal{K}_r$ in $\mathcal{M}$ and let $\nu \in \mathcal{M} \setminus \mathcal{V}$. As in the proof of Theorem 4.2.9 one shows that ν can be separated from $\mathcal{K}_r$ by some hyperplane, i.e., there exist $f_\nu \in C(X)$ and $\gamma_\nu > 0$ such that $\nu(f_\nu) > \gamma_\nu$ and $\tilde{\nu}(f_\nu) < 0$ for each $\tilde{\nu} \in \mathcal{K}_r$. Let $\mathcal{H}_\nu := \{\tilde{\nu} \in \mathcal{M} : \tilde{\nu}(f_\nu) > \gamma_\nu\}$ be the open half space containing ν. Then $\bar{\mathcal{H}}_\nu \subset \mathcal{M} \setminus \mathcal{K}_r$. Let $B_n := \{x : \varepsilon_{n,x} \in \mathcal{H}_\nu\}$. We are going to prove below that

$$\limsup_{n\to\infty} \frac{1}{n} \log m(B_n) \leq -r\,. \tag{6.4}$$

(This inequality corresponds to Proposition 5.6.3 in the Gibbs state setting.) Assuming this for the moment one can estimate

$$\begin{aligned}
&\limsup_{n\to\infty} \frac{1}{n} \log \mu\{x : \varepsilon_{n,x} \in \mathcal{M} \setminus \mathcal{V}\} \\
\leq\ & \max_{i=1,\dots,N} \limsup_{n\to\infty} \frac{1}{n} \log \mu\{x : \varepsilon_{n,x} \in \mathcal{H}_{\nu_i}\} \\
\leq\ & -r
\end{aligned}$$

with suitable $\nu_i \in \mathcal{M} \setminus \mathcal{V}$ just as in the proof of the first part of Theorem 5.6.1.

We turn to the proof of (6.4) omitting the subscript ν of f_ν and γ_ν. Recall that $B_n = \{x : \varepsilon_{n,x}(f) > \gamma\}$. We want to produce a measure $\mu \in \mathcal{M}(T)$ such that

$$\limsup_{n\to\infty} \frac{1}{n} \log m(B_n) \le F_\alpha(\mu) < -r\,.$$

As f is continuous and $\gamma > 0$, there is $\delta > 0$ such that $|f(x)-f(x')| < \gamma$ whenever $\rho(x,x') < \delta$. Fix any finite partition β with diam$(\beta) < \delta$ (see Lemma 4.4.10) and let $\mathcal{Z}_n := \{Z \in \alpha_0^n \vee \beta : Z \cap B_n \neq \emptyset\}$. For each $Z \in \mathcal{Z}_n$ fix a point $x'_Z \in Z \cap B_n$ and a point $x_Z \in Z$ with

$$m(T^n Z) = \int_Z e^{-\psi_n}\,dm \ge e^{-\psi_n(x_Z)} \cdot m(Z)$$

(see Remark 6.1.2-d). Let $E_n := \{x_Z : Z \in \mathcal{Z}_n\}$, $E'_n := \{x'_Z : Z \in \mathcal{Z}_n\}$. As $|E_n \cap A| \le \text{card}(\beta) < \infty$ for each $A \in \alpha_0^n$, it follows from Lemma 4.4.8 which prepared the construction of equilibrium states that there is a measure $\mu \in \mathcal{M}(T)$ with

$$F_\alpha(\mu) = h(\mu,\alpha,T) + \mu(\psi) \ge \limsup_{n\to\infty} \frac{1}{n} P(\psi_n, E_n)\,. \tag{6.5}$$

(For invoking this lemma we need assumption (6.2).) By construction, μ is the weak limit of a subsequence of the measures

$$\mu_n = \frac{1}{n}\sum_{k=0}^{n-1}\sum_{x\in E_n} e^{\psi_n(x)-P(\psi_n,E_n)} \cdot \delta_{T^k x} = \sum_{Z\in\mathcal{Z}_n} e^{\psi_n(x_Z)-P(\psi_n,E_n)} \cdot \varepsilon_{n,x_Z}\,.$$

We estimate $\varepsilon_{n,x_Z}(f)$. As $\rho(T^k x_Z, T^k x'_Z) \le \text{diam}(\beta) < \delta$ for all $k = 0,\dots,n-1$ we have

$$\varepsilon_{n,x_Z}(f) \ge \varepsilon_{n,x'_Z}(f) - \frac{1}{n}\sum_{k=0}^{n-1} |f(T^k x_Z) - f(T^k x'_Z)| > \varepsilon_{n,x'_Z}(f) - \gamma\,.$$

Now $E'_n \subseteq B_n$, i.e., $\varepsilon_{n,x'_Z}(f) > \gamma$. Therefore $\varepsilon_{n,x_Z}(f) \ge 0$. As each μ_n is a convex combination of measures ε_{n,x_Z} with $\varepsilon_{n,x}(f) \ge 0$, also $\mu(f) \ge 0$, in particular $\mu \notin \mathcal{K}_r$ so that $F_\alpha(\mu) < -r$.

Because of the choice of the x_Z we can estimate

$$m(B_n) \le \sum_{Z\in\mathcal{Z}_n} m(Z) \le \sum_{Z\in\mathcal{Z}_n} e^{\psi_n(x_Z)}\, m(T^n Z) \le \sum_{x\in E_n} e^{\psi_n(x)} = e^{P(\psi_n,E_n)}\,.$$

Combining this with (6.5) and the fact that $F_\alpha(\mu) < -r$ we obtain

$$\limsup_{n\to\infty} \frac{1}{n} \log m(B_n) \le \limsup_{n\to\infty} \frac{1}{n} P(\psi_n, E_n) \le F_\alpha(\mu) < -r\,,$$

i.e., (6.4). □

6.1.12 Corollary *$p(\psi) \geq 0$ under the assumptions of Theorem 6.1.8.*

Proof: Because of the theorem, $V_T(x) \subseteq \{\mu \in \mathcal{M}(T) : F(\mu) \geq 0\}$ for m-a.e. x. As $V_T(x) \neq \emptyset$ there is $\mu \in \mathcal{M}(T)$ with $F(\mu) \geq F_\alpha(\mu) \geq 0$. It follows that $p(\psi) = \sup\{F(\mu) : \mu \in \mathcal{M}(T)\} \geq 0$. □

In view of the corollary situations where $p(\psi) = 0$ are particularly interesting, because in those cases observable measures are equilibrium states for ψ. In the next sections we approach this question from various sides.

6.2 Transfer operators

Our next goal is to derive conditions on μ that imply $F(\mu) \leq 0$ and to find out when $p(\psi) = \sup_\mu F(\mu) \leq 0$. Combined with results from the previous section this will give us sufficient conditions for $F(\mu) = p(\psi) = 0$. As before suppose that $(X, \mathcal{B}, m, T, \alpha)$ is a fibred system. Given a function $\psi : X \to [-\infty, \infty]$ define the *transfer operator* $\mathcal{L}_\psi$ acting on the space of measurable functions $f : X \to [0, \infty]$ by

$$\mathcal{L}_\psi f = \sum_{A \in \alpha} (f\, e^\psi) \circ T_{|A}{}^{-1} .$$

Here and in the sequel we use the convention that $(\phi \circ T_{|A}^{-1})(x) = 0$ for $x \notin TA$ and any $\phi : X \to \mathbb{R}$.

6.2.1 Theorem (Free energy and the transfer operator)
Let $(X, \mathcal{B}, m, T, \alpha)$ be a fibred system. Consider $\mu \in \mathcal{M}(T)$ such that α is a (one-sided!) μ-generator for $\mathcal{B}$. Suppose there are measurable $\psi : X \to [-\infty, \infty]$ and $f : X \to [0, \infty]$ such that

1. *$\mu\{x : f(x) = 0$ or $f(x) = \infty\} = 0$,*
2. *$\psi^- \in L^1_\mu$ or $\psi^+ \in L^1_\mu$, i.e., $\mu(\psi)$ is well defined, and*
3. *$\mathcal{L}_\psi f \leq q \cdot f$ for some $q \in L^1_\mu$.*

Then $F(\mu) \leq \log \mu(q)$.

Proof: Suppose first that $\mu(\psi) = -\infty$. As $h(\mu) = h(\mu, \alpha, T) \leq \log(\text{card } \alpha) < \infty$, $F(\mu) = -\infty$ in this case, and nothing is to be proved. So we may assume that $\mu(\psi) > -\infty$, in particular $\mu(\psi^-) < \infty$.

Because of assumption 1, $\log f \circ T - \log f$ is μ-a.e. well defined. In the first step of the proof we show that it is also μ-integrable and that $\mu(\log f \circ T - \log f) = 0$. Let $q_1 = \max\{1, q\}$. Then $\log q_1 \in L^1_\mu$ and

$$f(x)\, e^{\psi(x)} \leq \sum_{A \in \alpha} (f\, e^\psi)(T_{|A}^{-1} T x) = (\mathcal{L}_\psi f)(Tx) \leq q_1(Tx)\, f(Tx)$$

for each x. As $\log q_1, \psi^- \in L^1_\mu$ and as the difference $\log f \circ T - \log f$ is μ-a.s. well defined,

$$\log f \circ T - \log f \geq \psi - \log q_1 \circ T \geq -\psi^- - \log q_1 \circ T$$

μ-a.s. and $\log f \circ T - \log f$ has a μ-integrable minorant. Therefore we can apply Lemma 4.1.13 and conclude that $\log f \circ T - \log f \in L^1_\mu$ and $\mu(\log f \circ T - \log f) = 0$.

We turn to the main part of the proof. As $\alpha_0^\infty = \mathcal{B} \bmod \mu$ by assumption, we have $\alpha_1^\infty = T^{-1}\mathcal{B} \bmod \mu$. In particular $\mu(A \mid \alpha_1^\infty) = \mu(A \mid T^{-1}\mathcal{B})$ for each $A \in \alpha$. As $\mu(A \mid T^{-1}\mathcal{B})$ is $T^{-1}\mathcal{B}$-measurable, there exists for each $A \in \alpha$ a $\mathcal{B}$-measurable function $p_A : X \to [0, 1]$ such that $p_A \circ T$ is a version of $\mu(A \mid \alpha_1^\infty)$. Let $\phi : X \to [0, \infty]$ be any $\mathcal{B}$-measurable function. As $p_A \circ T_{|A}^{-1}$ is $\mathcal{B}$-measurable we have

$$\begin{aligned}
\int_A \phi \, d\mu &= \int_A (\phi \circ T_{|A}{}^{-1}) \circ T \, d\mu = \int (p_A \circ T) \cdot (\phi \circ T_{|A}{}^{-1}) \circ T \, d\mu \\
&= \int p_A \cdot (\phi \circ T_{|A}{}^{-1}) \, d\mu
\end{aligned}$$

for each $A \in \alpha$. We apply this identity to $\phi = \frac{f \cdot e^\psi}{(p_A \cdot f) \circ T}$. By our considerations in the previous paragraph, $\frac{f}{f \circ T}$ is μ-a.e. well defined and, by Lemma 3.1.2, $p_A > 0$ μ-a.e. on A. Hence

$$\begin{aligned}
\int_A \frac{f \cdot e^\psi}{(p_A \cdot f) \circ T} \, d\mu &= \int p_A \cdot \frac{(f \cdot e^\psi) \circ T_{|A}{}^{-1}}{(p_A \cdot f) \circ T \circ T_{|A}{}^{-1}} \, d\mu \\
&= \int \frac{(f \cdot e^\psi) \circ T_{|A}{}^{-1}}{f} \, d\mu \, .
\end{aligned}$$

From this we obtain

$$\int_A e^{\psi - \log \mu(A|\alpha_1^\infty) - (\log f \circ T - \log f)} \, d\mu = \int \frac{(f \cdot e^\psi) \circ T_{|A}{}^{-1}}{f} \, d\mu \, .$$

Summing this identity over all $A \in \alpha$ and recalling assumption 3 lead to

$$\int e^{\psi - (\log f \circ T - \log f) + I_{\alpha|\alpha_1^\infty}} \, d\mu \leq \int \frac{\mathcal{L}_\psi f}{f} \, d\mu \leq \int q \, d\mu \, .$$

In view of Jensen's inequality (A.4.17) this implies

$$\mu(\psi - (\log f \circ T - \log f)) + \mu(I_{\alpha|\alpha_1^\infty}) \leq \log \mu(q) \, .$$

But $\mu(I_{\alpha|\alpha_1^\infty}) = H_\mu(\alpha \mid \alpha_1^\infty) = h(\mu, \alpha, T)$ by Theorem 3.2.7, and $h(\mu, \alpha, T) = h(\mu)$ as α is a μ-generator for $\mathcal{B}$. Above we saw that $\mu(\log f \circ T - \log f) = 0$. Hence $F(\mu) = \mu(\psi) + h(\mu) \leq \log \mu(q)$. □

6.2.2 Corollary *If, in the situation of the previous theorem, α is a generator, if $0 < f(x) < \infty$ for all $x \in X$, if $\sup \psi^- < \infty$ or $\sup \psi^+ < \infty$, and if $\mathcal{L}_\psi f \leq f$, then $p(\psi) \leq 0$.*

Proof: In this case, assumptions 1–3 of the theorem are satisfied with $q = 1$ for each $\mu \in \mathcal{M}(T)$, so that $p(\psi) = \sup_\mu F(\mu) \leq 0$. □

In the statement and in the proof of the theorem and its corollary the measure m itself played no role. This is changed in the following corollary where sufficient conditions are given that T has an invariant density:

6.2.3 Corollary *Let $(X, \mathcal{B}, m, T, \alpha)$ be an $e^{-\psi}$-conformal fibred system. Suppose there is a measurable $f : X \to [0, \infty]$ such that $\int f\, dm = 1$ and $\mathcal{L}_\psi f = f$ m-a.s.*

a) *Then $\mu := fm \in \mathcal{M}(T)$.*
b) *If $\alpha_0^\infty = \mathcal{B} \mod m$ (i.e., if α is a one-sided m-generator for T) and if $\psi^- \in L^1_\mu$ or $\psi^+ \in L^1_\mu$, then $F(\mu) \leq 0$.*

Proof:
a) $\mu = fm \in \mathcal{M}$, and for each measurable $\phi : X \to [0, \infty)$

$$\begin{aligned}\mu(\phi \circ T) &= \sum_{A \in \alpha} \int_A (\phi \circ T) \cdot f e^{\psi} \cdot e^{-\psi}\, dm = \sum_{A \in \alpha} \int_{TA} \phi \cdot (f e^{\psi}) \circ T_{|A}{}^{-1}\, dm \\ &= \int \phi \cdot \mathcal{L}_\psi f\, dm = \int \phi f\, dm \\ &= \mu(\phi)\,.\end{aligned}$$

Hence $\mu \in \mathcal{M}(T)$.

b) As under the additional hypothesis the assumptions of the theorem are satisfied with $q = 1$, $F(\mu) \leq 0$ in this case. □

Before we apply these results to the circle map examples from the previous section, we make some remarks on how the generator property of a partition α can be checked. As usual denote by $A_n(x)$ that element of α_0^n that contains x. Then we have

6.2.4 Lemma (Generating partitions for fibred systems) *Consider a fibred system $(X, \mathcal{B}, m, T, \alpha)$ where (X, ρ) is a metric space with Borel σ-algebra $\mathcal{B}$.*

a) *If $\lim_{n\to\infty} \operatorname{diam}(A_n(x)) = 0$ for each $x \in X$, then α is a generator. This condition is in particular met if some iterate of T is uniformly expanding, i.e., if there are $k \in \mathbb{N}$ and $\gamma > 1$ such that $\rho(T^k x, T^k y) > \gamma \cdot \rho(x, y)$ whenever x and y belong to the same element of α_0^k.*

b) Let $\nu \in \mathcal{M}$ and fix some $Y \in \mathcal{B}$ with $\nu(Y) = 1$. If $\lim_{n\to\infty} \operatorname{diam}(A_n(x) \cap Y) = 0$ for ν-a.e. $x \in X$, then α is a ν-generator.

Proof:

a) Let $U \subseteq X$ be open. For each $x \in X$ there exists an integer $n(x)$ such that $A_n(x) \subseteq U$. Therefore U is the countable (!) union of all elements $A \in \bigcup_{n>0} \alpha_0^n$ that are contained in U. In particular $U \in \alpha_0^\infty$, and it follows that $\mathcal{B} = \alpha_0^\infty$. If T^k is uniformly expanding, then $\rho(x,y) < \gamma^{-n}\rho(T^{nk}x, T^{nk}y) \leq \gamma^{-n} \operatorname{diam}(X)$ for all x and $y \in A_{nk}(x)$. Hence $\lim_{n\to\infty} \operatorname{diam}(A_{nk}(x)) = 0$.

b) Without loss of generality we can assume that $\lim_{n\to\infty} \operatorname{diam}(A_n(x) \cap Y) = 0$ for all $x \in Y$. (If necessary we delete all points x from Y for which this does not hold.) Then the same reasoning as before shows that there is $\tilde{U} \in \alpha_0^\infty$ such that $U \cap Y = \tilde{U} \cap Y$. Hence α is a ν-generator. □

6.2.5 Exercise Let $(X, \mathcal{B}, m, T, \alpha)$ be an $e^{-\psi}$-conformal fibred system. Show that $\int f \cdot (g \circ T)\, dm = \int \mathcal{L}_\psi f \cdot g\, dm$ for all bounded measurable $f, g : X \to \mathbb{R}$. ◇

6.2.6 Remark The preceding lemma applies to all examples in 6.1.4 except for the logistic maps. This is immediately clear if $|(T^k)'| > \gamma > 1$ uniformly for all branches of some iterate T^k.

In other cases let $\nu \in \mathcal{M}(T)$ and write $A_\infty(x) := \bigcap_n A_n(x)$. By definition, $TA_\infty(x) \subseteq A_\infty(Tx)$ so that $\nu(A_\infty(Tx)) \geq \nu(A_\infty(x))$. Hence $\nu(A_\infty(T^k x))$ is nondecreasing in k so that $\nu(A_\infty(x)) = 0$ or there are $k \geq 0$ and $l > 0$, both minimal, such that $T^l A_\infty(T^k x) \subseteq A_\infty(T^k x)$. Suppose that in the latter case $k > 0$. Then

$$\begin{aligned} \nu(A_\infty(T^k x)) &= \nu(T^{-1}A_\infty(T^k x)) \geq \nu(A_\infty(T^{k-1}x)) + \nu(T^{l-1}A_\infty(T^k x)) \\ &\geq \nu(A_\infty(T^{k-1}x)) + \nu(A_\infty(T^k x)) \,. \end{aligned}$$

Hence $\nu(A_\infty(x)) \leq \nu(A_\infty(T^{k-1}x)) = 0$. It follows that $T^l A_\infty(x) \subseteq A_\infty(x)$ if $\nu(A_\infty(x)) > 0$.

In the case of one-dimensional maps, $T^l_{|A_\infty(x)}$ is a strictly monotone continuous map of an interval. Therefore, if this map is increasing, ν is concentrated on the set of fixed points of T^l in $A_\infty(x)$. If ν is ergodic under T, it gives mass to at most one of these fixed points in $A_\infty(x)$ and we can choose the set Y in the preceding lemma such that $Y \cap A_\infty(x)$ contains just this one point. If T^l is decreasing on $A_\infty(x)$, we consider $T^{2l}_{|A_\infty(x)}$, which is increasing, and conclude that an ergodic ν can charge at most two points in $A_\infty(x)$, namely some points z and $T^l z$ where $T^{2l}z = z$. In such a situation we refine α so that these two points are separated and again we can manage to choose Y such that $Y \cap A_\infty(x)$ contains only one point. ◇

6.2.7 Example (Circle maps) We discuss the r-fold covering circle maps from Example 6.1.4-1a and 1b in the light of the previous results. Considered as maps on $[0,1)$ they have two monotone branches mapping the intervals $[0,\frac{1}{2})$ and $[\frac{1}{2},1)$ in a monotonically increasing way onto $[0,1)$. The point 0 (which is identified with the point 1) is a fixed point, and we are going to see that the slope of the map at this fixed point determines the nature of the dynamics. Because of Remark 6.1.10, Theorem 6.1.8 and Corollary 6.1.12 apply, so that $p(\psi) \geq 0$ and $F(\mu) \geq 0$ for each observable and for each absolutely continuous invariant measure μ.

Consider the map

$$T_\theta x = \frac{\theta x}{1+(\theta-2)x} \text{ if } 0 \leq x < \frac{1}{2}, \quad T_\theta x = 1 - T_\theta(1-x) \text{ if } \frac{1}{2} \leq x < 1$$

with $0 < \theta < \infty$. As the branches of T_θ, and hence also those of T_θ^2, are linear fractional, the monotonicity and symmetry properties of T'_θ and $(T_\theta^2)'$ imply that $T'_\theta(x)$ ranges between $T'_\theta(0) = \theta$ and $T'_\theta(\frac{1}{2}) = \frac{4}{\theta}$ and that $(T_\theta^2)'$ is minimized at the endpoints of the four monotone branches of T_θ^2. In fact,

$$\min_x (T_\theta^2)'(x) = \min\{\theta^2, (1+\frac{2}{\theta})^2\} > 1\,. \tag{6.6}$$

Let $\psi_\theta := -\log|T'_\theta|$. Then $\psi_\theta \geq \min\{-\log\theta, -\log\frac{4}{\theta}\} > -\infty$.

For $\theta > 1, \theta \neq 2$ let

$$f_\theta(x) := \frac{-1}{2\log(\theta-1)} \cdot \frac{\theta(\theta-2)}{(1-\theta+(\theta-2)x)(1+(\theta-2)x)}\,,$$

and set $f_1(x) := \frac{1}{x(1-x)}$, $f_2(x) := 1$. Observe that $0 < \inf_x f_\theta(x) \leq \sup_x f_\theta(x) < \infty$ for all $\theta > 1$ and that $0 < f_1(x) < \infty$ for $0 < x < 1$. It is not hard to check that $\mathcal{L}_{\psi_\theta} f_\theta = f_\theta$ for all $\theta \geq 1$.

$\theta > 1$: Then $\int f_\theta\, dm = 1$. Because of (6.6) the partition α is generating, so that Corollary 6.2.2 shows that $p(\psi_\theta) = 0$ and Corollary 6.2.3 that $\mu_\theta := f_\theta m \in ES(\psi_\theta, T_\theta)$ is an invariant measure equivalent to m. We are going to see in Example 6.3.6 that μ_θ is ergodic so that it is the only SBR measure for T_θ. In Corollary 6.3.10 we shall prove that each $\mu \in \mathcal{M}(T_\theta)$ which is absolutely continuous w.r.t. m is in $ES(\psi, T_\theta)$. Hence $ES(\psi, T_\theta) = \{\mu_\theta\}$.

$\theta = 1$: As $T'_1(x) > 1$ at all points x except at $x = 0$, it is easy to check that α is generating using Lemma 6.2.4-a. Hence Theorem 6.2.1 applies to all $\mu \in \mathcal{M}(T)$ except $\mu = \delta_0$. But $F(\delta_0) = -\log|T'_1(0)| = 0$, so that again $p(\psi_1) = 0$. In this case, however, $f_1 m$ is not a finite measure because $\int f_1\, dm = \infty$. But $\delta_0 \in ES(\psi_1, T_1)$. The reader is asked in Exercise 6.3.11 to show that $V_{T_1}(x) = \{\delta_0\}$ for m-a.e. x so that δ_0 is the only SBR measure for T_1.

$\theta < 1$: In this case a computation yields that

$$\mathcal{L}_{\psi_\theta} f_1(x) - f_1(x) = \frac{\theta^2 - 1}{(1-(1-\theta)x)(\theta + (1-\theta)x)} < 0$$

for all $x \neq 0$. As the partition α is μ-generating for all ergodic $\mu \in \mathcal{M}(T_\theta)$ by Remark 6.2.6, it follows from Theorem 6.2.1 that $F(\mu) < 0$ for all ergodic $\mu \in \mathcal{M}(T_\theta)$ with $\mu(\{0\}) = 0$. Therefore $F(\mu) = \int F(\mu_x)\, d\mu(x) < 0$ for all $\mu \in \mathcal{M}(T_\theta)$ with $\mu(\{0\}) < 1$ by Theorem 4.3.7. On the other hand, $F(\delta_0) = -\log |T'_\theta(0)| = -\log \theta > 0$. Hence $p(\psi) = -\log \theta$. It follows that $ES(\psi_\theta, T_\theta) = \{\delta_0\}$ and hence $V_{T_\theta} = \{\delta_0\}$ for m-a.e. x, i.e., δ_0 is the unique SBR measure of T_θ.

We turn to the map $Tx = 1 - \sqrt{1-2x}$ if $0 \leq x < \frac{1}{2}$, $Tx = 1 - T(1-x)$ if $\frac{1}{2} \leq x < 1$. Observe that $T'(0) = 1$ just as in the case of T_1 and that, as with that map, α is generating. However, in contrast to T_1, T has infinite slope at $\frac{1}{2}$, the second preimage of 0. This makes it possible that T has an integrable invariant density, namely $\mathcal{L}_\psi 1 = 1$ where again $\psi = -\log |T'|$. This means that the Lebesgue measure m itself is invariant, see Corollary 6.2.3. It follows from Theorem 6.1.8-c that $F(m) \geq 0$. One can show that m is ergodic, but the proof is beyond the scope of this text. The interested reader may consult [10]. In particular, $V_T(x) = \{m\}$ for m-a.e. x so that m is the only SBR measure for T.

As ψ is upper semicontinuous (although $\psi(\frac{1}{2}) = -\infty$) and as $\psi \leq 0$, Corollary 6.2.2 implies $F(m) = p(\psi) = 0$ and $m \in ES(\psi, T)$. However, there is one more equilibrium state for ψ, namely δ_0, because again $F(\delta_0) = -\log |T'(0)| = 0$. (Recall that we identify the points 0 and 1, i.e., $\delta_1 = \delta_0$.) So this is an example of a simple circle map which has two ergodic equilibrium states and where the topological support of one of them is contained in the topological support of the other. This situation is similar to that described in Remark 5.3.4. ◇

6.3 Absolutely continuous equilibrium states

In some of the examples of the last section we already encountered absolutely continuous invariant measures which turned out to be equilibrium states for $\psi = -\log |T'|$. In this section we show, under a suitable regularity assumption on the "geometry" of the branches of all iterates T^n, that the absolutely continuous invariant measures are exactly the equilibrium states. But before we turn to that question we study the ergodicity of such measures.

Unless otherwise stated we can relax the requirements for a fibred system in this section by allowing the partition α to be countable.

A central notion for $e^{-\psi}$-conformal fibred systems is the *distortion* of their branches. This is

$$D_\psi := \sup\{|\psi_n(x) - \psi_n(y)| : x, y \in A \in \alpha_0^n, n > 0\}.$$

6.3.1 Definition (Bounded distortion) *The transformation T has bounded distortion relative to m, if it is $e^{-\psi}$-conformal and if $D_\psi < \infty$.*

Recall that $e^{-\psi_n} = \prod_{k=0}^{n-1} e^{-\psi} \circ T^k$ is the "volume derivative" of T^n with respect to the measure m. Therefore T has bounded distortion if the "volume derivative" of all iterates of T is nearly constant (up to a factor e^{D_ψ}) along each single branch. Hence bounded distortion is equivalent to the existence of a constant $D'_\psi > 0$ such that

$$e^{\psi_n(x)-D'_\psi} \leq \frac{m(A_n(x))}{m(T^n A_n(x))} \leq e^{\psi_n(x)+D'_\psi} \tag{6.7}$$

for all $x \in X$ and all $n > 0$, because $m(T^n A_n(x)) = \int_{A_n(x)} e^{-\psi_n(y)}\,dm(y)$.

Distortion bounds are classically derived from two ingredients: T must be expanding in a suitable sense, and ψ must have good continuity properties. Here we assume that the $\psi_{|A}$ $(A \in \alpha)$ are uniformly *Hölder-continuous*. That means there exist $a \in (0,1]$ and $H > 0$ such that $|\psi(x) - \psi(y)| \leq H \cdot \rho(x,y)^a$ for all x, y belonging to the same element $A \in \alpha$.

6.3.2 Lemma (Distortion lemma) *Let $(X, \mathcal{B}, m, T, \alpha)$ be an $e^{-\psi}$-conformal fibred system where (X, ρ) is a metric space with Borel σ-algebra $\mathcal{B}$. Suppose that*

1. *ψ is Hölder-continuous,*
2. $\operatorname{diam}(\alpha) := \sup_{A\in\alpha} \operatorname{diam}(A) < \infty$ *and*
3. *there are $N \in \mathbb{N}$ and $\gamma > 1$ such that $\rho(T^N x, T^N y) \geq \gamma \cdot \rho(x,y)$ for all $A \in \alpha_0^N$ and $x, y \in A$.*

Then $(X, \mathcal{B}, m, T, \alpha)$ has bounded distortion.

Proof: Let $n > 0$, $A \in \alpha_0^n$ and $x, y \in A$. Decompose n as $n = rN + s$ with $r \in \mathbb{N}$ and $0 \leq s < N$. For $0 \leq k < r$ and $0 \leq i < N$ we have: $T^{(k-1)N+i}x$ and $T^{(k-1)N+i}y$ belong to the same element of α_0^N so that $\rho(T^{kN+i}x, T^{kN+i}y) \geq \gamma \cdot \rho(T^{(k-1)N+i}x, T^{(k-1)N+i}y)$. It follows by induction that

$$\operatorname{diam}(\alpha) \geq \rho(T^{(r-1)N+i}x, T^{(r-1)N+i}y) \geq \gamma^{r-k-1} \cdot \rho(T^{kN+i}x, T^{kN+i}y)\,. \tag{6.8}$$

Hence

$$\begin{aligned}
&|\psi_n(x) - \psi_n(y)| \\
&\leq \sum_{k=0}^{r-1}\sum_{i=0}^{N-1} |\psi(T^{kN+i}x) - \psi(T^{kN+i}y)| + \sum_{j=rN}^{n-1} |\psi(T^j x) - \psi(T^j y)| \\
&\leq \sum_{k=0}^{r-1}\sum_{i=0}^{N-1} H \cdot \rho(T^{kN+i}x, T^{kN+i}y)^a + s \cdot H \cdot \operatorname{diam}(\alpha)^a \\
&\leq \sum_{k=0}^{r-1}\sum_{i=0}^{N-1} H \cdot \gamma^{-a(r-k-1)} \cdot \operatorname{diam}(\alpha)^a + N \cdot H \cdot \operatorname{diam}(\alpha)^a \quad \text{by (6.8)}
\end{aligned}$$

$$\leq \; N \cdot H \cdot \operatorname{diam}(\alpha)^a \cdot \left(\frac{1}{1-\gamma^{-a}}+1\right) < \infty \,.$$

□

This lemma applies immediately to the expanding circle maps and the tent maps of Example 6.1.4.

6.3.3 Remark Bounded distortion plays a similar role as the regularity assumption (5.1.1) did in the study of Gibbs measures on shift spaces. Indeed, if the system $(X, \mathcal{B}, m, T, \alpha)$ is $e^{-\psi}$-conformal, then a slight modification of the previous proof shows that it has bounded distortion provided ψ is regular with respect to the sequence of partitions α_0^n. ◇

As a first application of distortion estimates we derive a condition which is sufficient for invariant measures $\mu \ll m$ of fibred systems to be ergodic. To this end recall that $\mathcal{I}(T) = \{B \in \mathcal{B} : T^{-1}B = B\}$.

6.3.4 Theorem (Ergodicity of fibred systems)
Let $(X, \mathcal{B}, m, T, \alpha)$ be a $e^{-\psi}$-conformal fibred system with bounded distortion. Suppose that $\alpha_0^\infty = \mathcal{B} \bmod m$ and that:

1. *For m-a.e. x there is some $U_x \in \mathcal{B}$ with $m(U_x) > 0$ for which there are $n_1(x) < n_2(x) < n_3(x) < \dots$ such that $U_x \subseteq T^{n_i(x)}(A_{n_i(x)}(x))$ for all $n_i(x)$.*
2. *There is some $U \in \mathcal{B}$ with $m(U) > 0$ such that $m(U \setminus \bigcup_{n=0}^{\infty} T^n U_x) = 0$ for m-a.e. x.*

Then $m(F) = 0$ or $m(F) = 1$ for each $F \in \mathcal{I}(T)$. (Although in general m is not T-invariant, we say that m is ergodic.)

6.3.5 Corollary *If, under the assumptions of the theorem, $\mu \in \mathcal{M}(T)$ is absolutely continuous w.r.t. m, then μ is ergodic.*

Proof: Let $F \in \mathcal{I}_\mu(T)$. There is $F' \in \mathcal{I}(T)$ such that $\mu(F \triangle F') = 0$ (see Remark 2.1.4). But $m(F') = 0$ or $m(F') = 1$ by the theorem, whence $\mu(F') = 0$ or $\mu(F') = 1$. □

Proof of the theorem: Let $F \in \mathcal{I}(T)$ and suppose for a contradiction that $0 < m(F) < 1$. Then

$$\frac{m(F \cap A_n(x))}{m(A_n(x))} = m(F \mid \alpha_0^n)(x) \to m(F \mid \alpha_0^\infty)(x) = 1_F(x)$$

for m-a.e. x by a corollary of the martingale convergence theorem (A.4.38). Since we assumed that $m(F) < 1$ it follows that

$$m\left\{x : \lim_{n\to\infty} \frac{m(F\cap A_n(x))}{m(A_n(x))} = 0\right\} > 0\,. \tag{6.9}$$

As $T^{-1}F = F$ we have

$$m(F\cap T^{n_i}A_{n_i}(x)) = m(T^{n_i}(F\cap A_{n_i}(x))) = \int_{F\cap A_{n_i}(x)} e^{-\psi_{n_i}}\,dm$$

for all $x \in X$ and all $n_i = n_i(x)$. Hence there is a set of points x of positive m-measure for which

$$\begin{aligned} m(F\cap U_x) &\le m(F\cap T^{n_i}A_{n_i}(x)) \le \frac{m(F\cap T^{n_i}A_{n_i}(x))}{m(T^{n_i}A_{n_i}(x))} \\ &= \frac{\int_{F\cap A_{n_i}(x)} e^{-\psi_{n_i}}\,dm}{\int_{A_{n_i}(x)} e^{-\psi_{n_i}}\,dm} \le e^{2D_\psi}\cdot\frac{m(F\cap A_{n_i}(x))}{m(A_{n_i}(x))} \\ &\to 0 \quad \text{as } i\to\infty. \end{aligned}$$

It follows that $m(F\cap U_x) = 0$ for these x and as T maps m-null sets to m-null sets,

$$m(F\cap U) \le m\left(F\cap \bigcup_{n=0}^{\infty} T^n U_x\right) = m\left(\bigcup_{n=0}^{\infty} T^n(F\cap U_x)\right) = 0\,.$$

As $0 < m(F) < 1$, the same argument can be applied to $X\setminus F$ so that also $m(U\setminus F) = 0$ which contradicts the hypothesis $m(U) > 0$. □

6.3.6 Example The measure $\mu = f_\theta m \in \mathcal{M}(T_\theta)$ for the circle map T_θ ($\theta > 1$) from Example 6.2.7 is ergodic, because it has bounded distortion by Lemma 6.3.2. ◇

6.3.7 Exercise Let T_s be a tent map with slope $s > 1$, see Example 6.1.4-2. Prove that Lebesgue measure is ergodic for T_s. *Hint:* There is $\delta > 0$ such that for each x there are infinitely many n with $m(T^n A_n(x)) > \delta$. Then a compactness argument yields the existence of intervals U_x needed in Theorem 6.3.4. Finally there is a one-sided neighbourhood U of $T_s(\frac{1}{2})$ such that for each interval J there is some $n > 0$ with $U \subseteq T^n J$. ◇

One further assumption on $(X, \mathcal{B}, m, T, \alpha)$ is formulated in the following theorem, which establishes a close link between absolute continuity w.r.t. m of measures $\mu \in \mathcal{M}(T)$ and the property that $F(\mu) = 0$. This assumption makes sure that high iterates of T do not have too short branches in the sense that $m(T^n A)$ for $A \in \alpha_0^n$ is not too small.

6.3.8 Theorem (Rohlin formula and Ruelle's inequality)
Let $(X, \mathcal{B}, m, T, \alpha)$ be an $e^{-\psi}$-conformal fibred system with bounded distortion and

$$\delta := \inf\{m(T^n A) : A \in \alpha_0^n, n > 0\} > 0 . \tag{6.10}$$

Consider $\mu \in \mathcal{M}(T)$ with $H_\mu(\alpha) < \infty$ and such that α is a μ-generator. Then:

a) $h(\mu) + \mu(\psi) \leq 0$. (This is a simple version of Ruelle's inequality.)

b) $h(\mu) + \mu(\psi) = 0$ if and only if $\mu \ll m$.

6.3.9 Remark Assuming the long branch condition (6.10) this theorem says that $\mu \ll m$ if and only if $h(\mu) + \int \psi \, d\mu = 0$, i.e., if μ satisfies the so called *Rohlin formula* $h(\mu) = \int \log JT \, d\mu$ where JT denotes the volume derivative of T. Unfortunately this assumption is violated in general. For r-fold covering circle maps, for the continued fraction transformation (see Example 6.3.16) and for the full tent map it is trivially satisfied, because for these maps $T^n A = X$ for all $A \in \alpha_0^n$. But already non-full tent maps as well as many higher-dimensional transformations need additional considerations to reduce the problem to a map that satisfies (6.10), see Example 6.3.13. ◇

Let $(X, \mathcal{B}, m, T, \alpha)$ be a *continuous* $e^{-\psi}$-conformal fibred system satisfying all assumptions of Theorem 6.3.8. Then the theorem can be interpreted in the setting of equilibrium state theory.

6.3.10 Corollary *Let $(X, \mathcal{B}, m, T, \alpha)$ be a continuous $e^{-\psi}$-conformal fibred system for which α is a finite generator satisfying condition (6.2). Then $p(\psi) = 0$, $ES(\psi, T) \neq \emptyset$, and*

$$\mu \in ES(\psi, T) \quad \Longleftrightarrow \quad h(\mu) = -\int \psi \, d\mu \quad \Longleftrightarrow \quad \mu \ll m .$$

Proof: $ES(\psi, T) \neq \emptyset$ by Theorem 4.2.3, $p(\psi) \geq 0$ by Corollary 6.1.12 and $p(\psi) = \sup_\mu(h(\mu) + \mu(\psi)) \leq 0$ because of Ruelle's inequality. Hence $\mu \in ES(\psi, T)$ if and only if $h(\mu) + \mu(\psi) = 0$, and this is equivalent to $\mu \ll m$ because of the theorem. □

We turn to the *proof of Theorem 6.3.8*, which is essentially taken from [15].

Let $\mu \in \mathcal{M}(T)$. As $\int_A e^{-\psi_n} \, dm = m(T^n A) \geq \delta > 0$ for all $A \in \alpha_0^n$ and all $n > 0$, and as $e^{-\psi} < \infty$ m-a.e. by Remark 6.1.2-b, it follows that also $m(A) > 0$ for all $A \in \alpha_0^n$ and all $n > 0$. We shall use this observation below.

We start with an estimate that is used in the proofs of both a) and b). Let

$$\beta_n := \{A \in \alpha_0^n : \mu(A) > 0\} \quad \text{and} \quad X_n := \sum_{A \in \beta_n} \frac{\mu(A)}{m(A)} \cdot 1_A .$$

Then

$$
\begin{aligned}
H_\mu(\alpha_0^n) &+ \int \psi_n\,d\mu \\
&= -\sum_{A\in\beta_n} \mu(A)\left(\log\mu(A) - \frac{1}{\mu(A)}\int_A \psi_n\,d\mu\right) \\
&\le -\sum_{A\in\beta_n} \mu(A)\Big(\log\mu(A) - \psi_n(x_A)\Big) \qquad \text{for suitable } x_A \in A, \\
&= -\sum_{A\in\beta_n} \mu(A)\log\Big(\mu(A)\cdot\exp(-\psi_n(x_A))\Big) \\
&\le D'_\psi - \sum_{A\in\beta_n} \mu(A)\log\left(\frac{\mu(A)}{m(A)}\cdot m(T^nA)\right) \qquad \text{because of (6.7),} \\
&= D'_\psi - \sum_{A\in\beta_n} m(A)\frac{\mu(A)}{m(A)}\log\frac{\mu(A)}{m(A)} - \sum_{A\in\beta_n}\mu(A)\log m(T^nA) \\
&\le D'_\psi + \int \varphi(X_n)\,dm - \log\delta\,,
\end{aligned}
$$

where again $\varphi(t) = -t\log t \le e^{-1}$.

a) As $h(\mu,\alpha,T) + \mu(\psi) = h(\mu) + \mu(\psi)$, it follows that

$$
h(\mu) + \mu(\psi) = \lim_{n\to\infty}\frac{1}{n}\left(H_\mu(\alpha_0^n) + \mu(\psi_n)\right) \le \lim_{n\to\infty}\frac{1}{n}\left(D'_\psi + e^{-1} - \log\delta\right) = 0\,.
$$

This proves assertion a).

b) Suppose first that $h(\mu) + \mu(\psi) \ge 0$. Then

$$
0 \le n(h(\mu) + \mu(\psi)) \le H_\mu(\alpha_0^n) + \mu(\psi_n) \le D'_\psi + \int \varphi(X_n)\,dm - \log\delta \qquad (6.11)
$$

by Theorem 3.2.4. Below we show that this inequality implies the uniform integrability of the random variables X_n with respect to the probability measure m. Before that we use this fact to prove that $\mu \ll m$.

Denote by $\mathcal{F}_n$ the σ-algebra generated by α_0^n. It is easy to see that $(X_n, \mathcal{F}_n)_{n>0}$ is a martingale. Because of its uniform integrability it converges in L^1_m to a random variable X_∞ and $E_m[X_\infty \mid \mathcal{F}_n] = X_n$ for each $n > 0$, see A.4.37. In particular, $\mu(A) = \int_A X_n\,dm = \int_A X_\infty\,dm$ for each $A \in \alpha_0^n$ and $n > 0$ so that $X_\infty = \frac{d\mu}{dm}$ (see A.4.7), whence $\mu \ll m$.

We show that the X_n are uniformly integrable: Because of (6.11) we have for each $M > 1$

$$
-\int_{\{X_n\ge M\}} \varphi(X_n)\,dm = -\int \varphi(X_n)\,dm + \int_{\{X_n<M\}} \varphi(X_n)\,dm \le D'_\psi - \log\delta + \frac{1}{e}\,,
$$

so that

$$\begin{aligned}\int_{\{X_n \geq M\}} X_n \, dm &\leq \int_{\{X_n \geq M\}} X_n \cdot \frac{\log X_n}{\log M} \, dm \\ &= -\frac{1}{\log M} \int_{\{X_n \geq M\}} \varphi(X_n) \, dm \leq \frac{D'_\psi - \log \delta + \frac{1}{e}}{\log M}\end{aligned}$$

for all $n > 0$. Therefore

$$\lim_{M \to \infty} \sup_{n > 0} \int_{\{X_n \geq M\}} X_n \, dm = 0 \, ,$$

which is one of the characterizations of uniform integrability (see A.4.21).

It remains to prove that $h(\mu) + \mu(\psi) \geq 0$ if $\mu \ll m$. For continuous systems for which the $A \in \alpha_0^n$ have "small" boundaries in the sense of condition (6.2) this is part of Theorem 6.1.8. Here we give a proof that does not rely on a topological framework: Suppose first that μ is ergodic. As $\mu \ll m$, the quotient $\frac{\mu(A_n(x))}{m(A_n(x))}$ converges μ-a.s. to some finite strictly positive limit, namely to $\frac{d\mu}{dm}$ (see A.4.39). It follows that

$$\limsup_{n \to \infty} \frac{1}{n} \log \frac{\mu(A_n(x))}{m(A_n(x))} = 0 \tag{6.12}$$

for μ-a.e. x. As $\psi(x) \leq D'_\psi - \log m(TA_1(x))$ for μ-a.e. x by (6.7) so that $\mu(\psi^+) \leq D'_\psi - \log \delta < \infty$ by (6.10), Lemma 2.2.2 implies that

$$\lim_{n \to \infty} \frac{1}{n} \psi_n(x) = \mu(\psi)$$

for μ-a.e. x. Finally, by the Shannon–McMillan–Breiman theorem 3.2.7

$$\lim_{n \to \infty} \frac{1}{n} \log \mu(A_n(x)) = -h(\mu, \alpha, T)$$

for μ-a.e. x. Combining these three limits we obtain

$$-\lim_{n \to \infty} \frac{1}{n} \log \left(m(A_n(x)) \cdot e^{-\psi_n(x)} \right) = h(\mu, \alpha, T) + \mu(\psi)$$

for μ-a.e. x and, observing (6.7) and the "long branch" assumption (6.10), we conclude that

$$-\lim_{n \to \infty} \frac{1}{n} \log m(T^n A_n(x)) = h(\mu, \alpha, T) + \mu(\psi)$$

for μ-a.e. x. As α is a μ-generator, this finishes the proof for ergodic μ.

If μ is not ergodic, we apply the above consideration to its ergodic components μ_x, and the result follows in view of Theorem 3.3.2. □

Theorem 6.3.8 and Corollary 6.3.10 furnish the last pieces of information to complete the discussion of circle maps in Example 6.2.7, except for the observation that $V_{T_1}(x) = \{\delta_0\} = ES(\psi_1, T_1)$ for m-a.e. x, which the reader is asked to prove in the following exercise.

6.3.11 Exercise Consider the circle map $T_1(x) = \frac{x}{1-x}$ if $0 \leq x < \frac{1}{2}$, $T_1(x) = 1 - T_1(1-x)$ if $\frac{1}{2} \leq x < 1$.

a) For $N \in \mathbb{N}$, $N > 0$, let $A_N := [\frac{1}{N}, \frac{N-1}{N}]$. Prove that for each $x \in (0,1)$ there exists $n \geq 1$ such that $T_1^n(x) \in A_N$. Hence the first return time τ_{A_N} of T_1 to A_N is finite for all $x \in (0,1)$.
b) Show that all branches of T_{A_N} have $[\frac{1}{N-1}, \frac{N-1}{N}]$, $[\frac{1}{N}, \frac{N-1}{N}]$, or $[\frac{1}{N}, \frac{N-2}{N-1}]$ as their ranges.
c) Denote by T_{A_N} the first return map to A_N, see Section 2.4, and let $f_{A_N}(x) := \frac{1}{2\log(N-1)} \frac{1}{x(1-x)}$. Prove that $\mu_{A_N} := f_{A_N} m$ is an ergodic invariant measure for T_{A_N} and that $\mu_{A_N}(\tau_{A_N}) = \infty$.
d) Use Birkhoff's ergodic theorem to show that $\nu([\frac{1}{N}, \frac{N-1}{N}]) = 0$ for each $\nu \in V_{T_1}(x)$ and μ_{A_N}-a.e. x. Conclude that $V_{T_1}(x) = \{\delta_0\}$ for m-a.e. x. (Recall that the points 0 and 1 are identified on the circle.)
e) Let $\nu \in \mathcal{M}(T)$ be ergodic, $\nu \neq \delta_0$. Prove that $h(\nu) + \nu(\psi) < 0$. *Hint:* Consider the normalized restriction ν_{A_N} of ν to A_N, see Section 2.4. ◇

6.3.12 Example (r-fold covering maps) For r-fold covering circle maps and for the full tent map we have $T^n A = X$ for all $A \in \alpha_0^n$ (we say that T has *full branches*.) In particular, the long branch condition (6.10) is satisfied. We can summarize our results for r-fold covering strictly expanding circle maps as in Example 6.1.4-1 (and their analogues on higher-dimensional tori) as follows:

Let T be a strictly expanding r-fold covering C^2 map of the circle or of a higher-dimensional torus. Then

$$\begin{aligned} & \mu \text{ is an equilibrium state for } -\log|\det DT| \\ \text{if and only if } & T \text{ satisfies the Rohlin formula } h(\mu) = \int \log|\det DT|\, d\mu \\ \text{if and only if } & \mu \ll m. \end{aligned}$$

The only additional problem for the higher-dimensional case is to verify assumption (6.2) of Theorem 6.1.8. A reader who wants to do this should observe that for these maps we are fairly free to choose our partition α. ◇

6.3.13 Example (Symmetric tent maps) Recall that $T_s : [0,1] \to [0,1]$, $T_s x = s \cdot (\frac{1}{2} - |x - \frac{1}{2}|)$ with $0 < s \leq 2$. We note that $|T_s'| = s$, i.e., $\psi_s := -\log|T_s'| = -\log s$ is constant, so that $F(\mu) = h(\mu) - \log s$ for all $\mu \in \mathcal{M}(T)$, and $\mu \in ES(\psi_s, T_s)$ if and only if μ is a measure of maximal entropy.

$s < 1$: In this case T_s is a contraction on $[0, 1]$ with fixed point $x = 0$. Hence $\mathcal{M}(T_s) = \{\delta_0\}$ so that $V_{T_s}(x) = \{\delta_0\} = ES(\psi_s, T_s)$ for all x and $p(\psi) = F(\delta_0) = -\log s > 0$.

$s = 1$: As $T_{1|[0,\frac{1}{2}]}$ is the identity on $[0, \frac{1}{2}]$ and as $T_1[\frac{1}{2}, 1] = [0, \frac{1}{2}]$, we have $\mathcal{M}(T_1) = \{\mu \in \mathcal{M} : \mu([0, \frac{1}{2}]) = 1\}$ and $F(\mu) = 0$ for all $\mu \in \mathcal{M}(T)$. Hence $p(\psi_s) = 0$, $V_{T_1}(x) = V_{T_1}(1-x) = \{\delta_x\}$ for all $x \in [0, \frac{1}{2}]$ and $ES(\psi_1, T_1) = \mathcal{M}(T)$.

$s > 1$: The only obstacle to applying Corollary 6.3.10 in this case is that T_s does not have full branches for $s < 2$ and that in general $\inf\{m(T^n A) : A \in \alpha_0^n, n > 0\} = 0$. We shall indicate below how this problem can be overcome and suppose for the moment that the conclusions of Corollary 6.3.10 hold. Then $p(\psi_s) = 0$, the absolutely continuous invariant measures are just the equilibrium states for ψ_s, and as $ES(\psi_s, T_s) \neq \emptyset$, such a measure exists. Because of Theorem 6.1.8, $V_T(x) \subseteq ES(\psi_s, T_s)$ for m-a.e. x. The reader was asked in Exercise 6.3.7 to show that Lebesgue measure is ergodic for T_s. This shows that there is only one $\mu \in \mathcal{M}(T)$ which is absolutely continuous w.r.t. m, so $ES(\psi_s, T_s) = \{\mu\}$.

We indicate how the length of the branches of T_s^n for $s > 1$ can be controlled: With a map $T = T_s$ one can associate a conformal fibred system $(\hat{X}, \hat{\mathcal{B}}, \hat{m}, \hat{T}, \hat{\alpha})$ with countably many branches (called the Markov extension of T_s) in such a way that

1. this system is s-conformal (just as T_s is),
2. there is a one-to-one correspondence between measures $\mu \in \mathcal{M}(T)$ and measures $\hat{\mu} \in \mathcal{M}(\hat{T})$ such that $h(\mu) = h_{\hat{T}}(\hat{\mu})$,
3. $\mu \ll m$ if and only if $\hat{\mu} \ll \hat{m}$,
4. μ is ergodic if and only if $\hat{\mu}$ is ergodic, and
5. $\hat{T}^{n-1}\hat{A} \in \hat{\alpha}$ for each $\hat{A} \in \hat{\alpha}_0^n$. (This is the new feature of the Markov extension.)

Hence the conclusions of Corollary 6.3.10 hold for T_s if and only if they hold for its Markov extension. A rather general version of this construction, which is based on entropy techniques, can be found in [33], see also [11].

Let $\mu \in \mathcal{M}(T)$ be ergodic. In the extension one fixes an $\hat{A} \in \hat{\alpha}$ with $\hat{\mu}(\hat{A}) > 0$. Denote by $\hat{\mu}_{\hat{A}}$ the normalized restriction of $\hat{\mu}$ to $\hat{A}$. Then $\hat{\mu} \ll \hat{m}$ if and only if $\hat{\mu}_{\hat{A}} \ll \hat{m}$, $\hat{T}_{\hat{A}}$ is $s^{\tau_{\hat{A}}}$-conformal and

▷ $\int \log s \, d\hat{\mu} = \int_{\hat{A}} \sum_{k=0}^{\tau_{\hat{A}}-1} \log s \, d\hat{\mu} = \hat{\mu}(\hat{A}) \cdot \int \log s^{\tau_{\hat{A}}} \, d\hat{\mu}_{\hat{A}}$ by Exercise 2.4.5,

▷ $h_{\hat{T}}(\hat{\mu}) = \hat{\mu}(\hat{A}) \cdot h_{\hat{T}_{\hat{A}}}(\hat{\mu}_{\hat{A}})$ by Exercise 3.2.23.

Hence $h_{\hat{T}}(\hat{\mu}) = (\geq) \int \log s \, d\hat{\mu}$ if and only if $h_{\hat{T}_{\hat{A}}}(\hat{\mu}_{\hat{A}}) = (\geq) \int \log s^{\tau_{\hat{A}}} \, d\hat{\mu}_{\hat{A}}$, and the conclusions of Theorem 6.3.8 for $\hat{T}$ will follow from those for $\hat{T}_{\hat{A}}$. But $\hat{T}_{\hat{A}}$ clearly gives rise to a fibred system with countably many branches, all of them full, and with distortion zero, so that Theorem 6.3.8 applies to this map. By the argument in the previous paragraph, these conclusions also hold for the original T_s. The ergodicity of T can be reduced in the same way to that of $\hat{T}_{\hat{A}}$. ◇

6.3.14 Example (The full logistic map) For the logistic map $T_a(x) = ax(1-x)$ on $[0,1]$ the function $\psi(x) = -\log|T_a'(x)| = -\log a - \log|1-2x|$ takes the value $+\infty$ at $x = \frac{1}{2}$. So there is no chance to apply the general variational formalism to this map, see [9] for a detailed discussion of this problem. Also T_a does not have bounded distortion. Nevertheless the case $a = 4$ of a full logistic map can be treated by a simple trick: Let $f(x) = \frac{1}{\pi\sqrt{x(1-x)}}$ for $0 \leq x \leq 1$. A simple calculation yields that $|T_4'(x)| = 2\frac{f(x)}{f(T_4 x)}$ so that $\psi_4(x) = -\log 2 + \log f(T_4 x) - \log f(x)$. As $\frac{f(T_4 x)}{f(x)} = \frac{2}{|T_4'(x)|} = \frac{1}{2|1-2x|} \geq \frac{1}{2}$ for $x \in [0,1]$, Lemma 4.1.13 implies that $\mu(\log f \circ T_4 - \log f) = 0$ for all $\mu \in \mathcal{M}(T_4)$. Hence $F(\mu) = h(\mu) - \log 2$ for all μ, in particular $ES(\psi_4, T_4) = ES(-\log 2, T_4)$. In fact, it turns out that $\mathcal{L}_{\psi_4} f = f$ so that $fm \in \mathcal{M}(T_4)$ and $F(fm) = 0$ by Corollary 6.2.3 and Theorem 6.1.8. ◇

6.3.15 Exercise Let $\Phi : [0,1] \to [0,1], x \mapsto \int_0^x f(t)\,dt$ with f as in the previous example. Show that $T_4(x) = \Phi^{-1} \circ S \circ \Phi(x)$, where S denotes the full tent map with slope $s = 2$. ◇

We finish this section with a classical example of a piecewise monotonic interval map with countably many branches.

6.3.16 Example (Continued fraction transformation) Let $T_{cf} : (0,1] \to (0,1]$, $T_{cf}x = \frac{1}{x} \bmod 1$. As $|(T_{cf}^2)'| \geq 2$, the fibred system is generating and has bounded distortion, see Lemmas 6.2.4 and 6.3.2. Obviously T_{cf} has only full branches. Thus a T-invariant measure μ with $H_\mu(\alpha) < \infty$ is absolutely continuous w.r.t. Lebesgue measure if and only if $F(\mu) = 0$, i.e., if $h_T(\mu) = 2 \cdot \int_0^1 |\log x| \, d\mu(x)$.

The T-invariance of the measure $d\mu_G(x) := \frac{1}{\log 2}\frac{dx}{1+x}$ was already known to Gauss [22]. It follows easily from Theorem 6.2.1. By Corollary 6.3.5 this measure is ergodic, in particular $V_{T_{cf}} = \{\mu_G\}$ for m-a.e. x. In order to check that $H_{\mu_G}(\alpha) < \infty$ just observe that $\alpha = ((\frac{1}{k+1}, \frac{1}{k}] : k = 1, 2, \ldots)$ and that $\mu_G(\frac{1}{k+1}, \frac{1}{k}] \leq \frac{1}{\log 2 \cdot (k+1)^2}$. Therefore Theorem 6.3.8 implies that $0 = F(\mu_G) \geq F(\nu)$ for all $\nu \in \mathcal{M}(T)$ with equality if and only if $\nu = \mu_G$. In particular $h(\mu_G) = \frac{2}{\log 2} \cdot \int_0^1 \frac{|\log x|}{1+x} \, dx = \frac{\pi^2}{6 \log 2}$. As T_{cf} cannot be turned into a continuous map of a compact metric space, we do not call μ_G an equilibrium state for ψ, although it maximizes the functional F. ◇

6.4 Iterated function systems (IFS)

Fractal sets have received great attention during the last two decades. One of the most popular ways to generate and explore a variety of them is by using *iterated function systems*. By this we mean the following:

6.4.1 Definition *Let $U \subset \mathbb{R}^d$ be a convex bounded open set. An iterated function system (IFS) is a finite family $\mathcal{F} = (f_\sigma : \sigma \in \Sigma)$, $|\Sigma| \geq 2$, of contracting self maps of U with the following properties:*

1. *There is $a < 1$ such that $|f_\sigma(x) - f_\sigma(y)| \leq a \cdot |x - y|$ for all $\sigma \in \Sigma$ and $x, y \in U$. (Here $|\,.\,|$ denotes the Euclidean norm on $\mathbb{R}^d$.)*
2. *There is a compact, convex subset $K \subseteq U$ such that $f_\sigma(K) \subseteq K$ for all $\sigma \in \Sigma$.*

◇

Basic facts on fractals generated by iterated function systems and many examples are presented in [1]. Most of the mathematics around this topic can be found in [19].

There are several ways to define the fractal set associated with $\mathcal{F}$. We use the most "dynamical" one that also fits best into our ergodic theoretic framework. For $v \in \Sigma^n$ define $f(v, .) : U \to U$, $f(v, z) := f_{v_n} \circ \cdots \circ f_{v_1}(z)$. Let $\Omega := \Sigma^{\mathbb{Z}}$ and for $\omega \in \Omega$ and integers $k < l$ let $\omega_k^l := \omega_{k+1} \ldots \omega_l$. We are going to define a map $F : \Omega \to U$ by

$$F(\omega) := \lim_{n \to \infty} f(\omega_{-n}^0, z), \quad z \in U \text{ arbitrary.}$$

The next lemma shows that F is well defined. The set $F(\Omega)$ is the fractal set we are looking for. Observe that F depends on ω only through $\omega_{-\infty}^0$ and that

$$F(T^k\omega) = f(\omega_0^k, F\omega) \quad \text{and} \quad f(\omega_{-k}^0, F(T^{-k}\omega)) = F(\omega) \tag{6.13}$$

for all $\omega \in \Omega$ and all $k \in \mathbb{Z}_+$ where T denotes the left shift on Ω.

6.4.2 Lemma *Let $\mathcal{F}$ be as above. Then*

$$|f(v, x) - f(v, y)| \leq a^n \cdot |x - y| \tag{6.14}$$

for all $v \in \Sigma^n$ and $x, y \in U$, and

$$\lim_{n \to \infty} f(\omega_{-n}^0, z) \text{ exists and is independent of } z \tag{6.15}$$

for all $\omega \in \Omega$ and $z \in U$. Finally, if $\omega, \tilde{\omega} \in \Omega$ are such that $\omega_{-n}^0 = \tilde{\omega}_{-n}^0$, then

$$|F(\omega) - F(\tilde{\omega})| \leq a^n \cdot \operatorname{diam}(U)\,, \tag{6.16}$$

i.e., F is Hölder-continuous. In particular, $F(\Omega)$ is compact.

Proof: The first claim follows by induction from the uniform contraction property of the system $\mathcal{F}$. For the second claim fix $z \in U$. Then

$$\begin{aligned}|f(\omega^0_{-(n+m)}, z) - f(\omega^0_{-n}, z)| &= |f(\omega^0_{-n}, f(\omega^{-n}_{-(n+m)}, z)) - f(\omega^0_{-n}, z)| \\ &\leq a^n \cdot |f(\omega^{-n}_{-(n+m)}, z) - z| \\ &\leq a^n \cdot \mathrm{diam}(U)\end{aligned}$$

for all $m, n > 0$ by (6.14). Therefore $(f(\omega^0_{-n}, z))_{n>0}$ is a Cauchy sequence and thus converges to $F(\omega)$. The independence of $F(\omega)$ from z follows again from (6.14). Estimate (6.16) is a consequence of (6.14) when this is applied with $x = F(T^{-n}\omega)$ and $y = F(T^{-n}\tilde{\omega})$. □

The following corollary expresses the "self-similarity" of $F(\Omega)$. (Strictly speaking, $F(\Omega)$ is self-similar only if all f_σ are similarity transformations.)

6.4.3 Corollary *$F(\Omega) \subseteq K$ and $F(\Omega) = \bigcup_{v \in \Sigma^n} f(v, F(\Omega))$ for all $n > 0$.*

6.4.4 Example (Sierpinski gasket) We consider an IFS that generates a fractal set which is a slight variation of the so called Sierpinski gasket. Let $\Sigma = \{0, 1, 2\}$ and let K be the equilateral triangle with corners $A_\sigma = e^{2\pi i \frac{\sigma}{3}}$ in the complex plane ($\sigma \in \Sigma$). Denote by f_σ the linear contraction with centre A_σ and contraction rate a_σ. If $U = B_r(0)$ with $r \geq \frac{1}{1-a_\sigma}$ for all $\sigma \in \Sigma$, then U is a convex, bounded, open neighbourhood of K, $f_\sigma(U) \subseteq U$ and $f_\sigma(K) \subseteq K$ for all $\sigma \in \Sigma$. Figure 6.1 shows a sketch of the set $F(\Omega)$ for $a_0 = 0.6$, $a_1 = 0.4$ and $a_2 = 0.3$. ◇

Figure 6.1: The fractal set from Example 6.4.4.

The naive way to generate an approximate picture of the compact set $F(\Omega)$ would be to choose n so large that a^n is of the order of the resolution of one's graphical device and then to generate systematically all points $f(v, z)$, $v \in \Sigma^n$, for some fixed $z \in U$. But already for a "tame" situation like $|\Sigma| = 3$, $a = 0.6$ and a resolution of 10^{-3} this would require one to compute about $3 \cdot 10^7$ points, while at the same time the graphical device would only allow one to represent 10^6 different points. Therefore the following random algorithm, which was also used to generate Figure 6.1, is a useful alternative, see also [1, Section III.8]

Let μ be any ergodic T-invariant probability measure on Ω with full topological support. (The topological support of μ is the smallest closed set of full measure; so full topological support means that each open set in Ω has positive μ-measure.) One of the implications of the next theorem, which is a rather direct consequence of Birkhoff's ergodic theorem, is that μ-typical random trajectories $(f(\omega_0^k, x))_{k>0}$

of the IFS $\mathcal{F}$ are dense in $F(\Omega)$. See also [1, Section IX.7]. For brevity we use the notations $F\Omega := F(\Omega)$ and $F\mu := \mu \circ F^{-1}$.

6.4.5 Theorem (Ergodic theorem for IFS)
Let $\mu \in \mathcal{M}(T)$ be ergodic. For each $x \in U$ and μ-a.e. $\omega \in \Omega$

$$\lim_{n\to\infty} \frac{1}{n} \sum_{k=1}^{n} \delta_{f(\omega_0^k, x)} = F\mu \quad \textit{(weak limit)}$$

where δ_y denotes as usual the unit point mass in y.

From this the denseness of typical random trajectories follows because $F\mu$ has full support in $F\Omega$.

Proof of the theorem: Let $\phi \in C(\bar{U})$. By Birkhoff's ergodic theorem

$$\lim_{n\to\infty} \frac{1}{n} \sum_{k=1}^{n} \phi(F(T^k\omega)) = \int \phi \circ F \, d\mu \quad \text{for } \mu\text{-a.e. } \omega. \tag{6.17}$$

As $F(T^k\omega) = f(\omega_0^k, F\omega)$ by (6.13), we have

$$|F(T^k\omega) - f(\omega_0^k, x)| \le a^k \cdot \text{diam}(U)$$

for all $x \in U$. As ϕ is continuous and as there is a countable dense subset of functions $\phi \in C(\bar{U})$, the lemma follows from (6.17). □

6.5 Pressure and dimension for IFS

The ergodic theorem for IFS tells us that $F\mu$ is the asymptotic distribution of a typical random trajectory on $F\Omega$. But some measures μ may yield very unevenly distributed trajectories, others may behave better in this respect, see Figure 6.2 on p. 151. So the next step is to single out those *measures* $F\mu$ which represent the *set* $F\Omega$ as accurately as possible. Our criterion for such a measure is that it is not concentrated on a measurable subset of $F\Omega$ that has smaller Hausdorff dimension than $F\Omega$ itself.

We proceed with the definition of the *Hausdorff dimension* of a set $A \subset \mathbb{R}^d$. For $\epsilon > 0$ let $\mathcal{U}_\epsilon(A)$ be the collection of all finite or countable coverings $(U_n)_n$ of A by balls of diameter at most ϵ. For $\epsilon > 0$ and $t > 0$ let

$$\begin{aligned} \mathcal{H}_\epsilon^t(A) &:= \inf\left\{ \sum_n (\text{diam}(U_n))^t : (U_n)_n \in \mathcal{U}_\epsilon(A) \right\}, \\ \mathcal{H}^t(A) &:= \sup_{\epsilon>0} \mathcal{H}_\epsilon^t(A) = \lim_{\epsilon\to 0} \mathcal{H}_\epsilon^t(A). \end{aligned}$$

If $t < t'$, then $\mathcal{H}^{t'}_\epsilon(A) \le \epsilon^{t'-t}\mathcal{H}^t_\epsilon(A)$, whence $\mathcal{H}^t(A) = \infty$ if $\mathcal{H}^{t'}(A) > 0$ and $\mathcal{H}^{t'}(A) = 0$ if $\mathcal{H}^t(A) < \infty$. Therefore there is a unique critical t, denoted by $HD(A)$, such that

$$\mathcal{H}^t(A) = \begin{cases} \infty & \text{if } t < HD(A), \\ 0 & \text{if } t > HD(A). \end{cases}$$

$HD(A)$ is called the *Hausdorff dimension* of A. Derived from this is the notion of *Hausdorff dimension of a probability measure* ν on $\mathbb{R}^d$:

$$HD(\nu) := \inf\{HD(A) : A \subseteq \mathbb{R}^d \text{ Borel measurable}, \nu(A) = 1\}.$$

6.5.1 Remark In the definition of $\mathcal{H}^t_\epsilon(A)$ it is more common to use coverings by any sets U_n with diameter at most ϵ, but this changes the value of $\mathcal{H}^t_\epsilon(A)$ by not more than a factor 2^t. With this convention the quantity $\mathcal{H}^t(A)$ is called the t-dimensional *Hausdorff measure* of A. For a systematic treatment of pressure and dimension in a unifying framework see [45]. ◇

For later use we note an important property of the function $A \mapsto HD(A)$:

6.5.2 Lemma *Let $A_j \subseteq X$, $j \in \mathbb{N}$. Then $HD(\bigcup_j A_j) = \sup_j HD(A_j)$.*

Proof: As $A_i \subseteq \bigcup_j A_j$ for all i, we have $HD(A_i) \le HD(\bigcup_j A_j)$ for all i so that $HD(\bigcup_j A_j) \ge \sup_i HD(A_i)$. For the reverse inequality assume that $t > \sup_j HD(A_j)$ and $\epsilon > 0$. Then $\mathcal{H}^t(A_j) = 0$ for all j and there are coverings $(U^j_n)_{n\in\mathbb{N}}$ of the A_j with $\operatorname{diam}(U^j_n) \le \epsilon$ such that $\sum_n(\operatorname{diam}(U^j_n))^t < 2^{-j}$. It follows that $(U^j_n)_{n,j\in\mathbb{N}}$ is a covering of $\bigcup_j A_j$ with $\sum_{n,j}(\operatorname{diam}(U^j_n))^t < \sum_{j\in\mathbb{N}} 2^{-j} = 2$. Hence $\mathcal{H}^t_\epsilon(\bigcup_j A_j) \le 2$ for all $\epsilon > 0$ so that $\mathcal{H}^t(\bigcup_j A_j) \le 2$ and $HD(\bigcup_j A_j) \le t$. As $t > \sup_j HD(A_j)$ was arbitrary, this finishes the proof of the lemma. □

We restrict our attention to *nondegenerate* IFS $\mathcal{F}$, i.e., to such $\mathcal{F}$ for which all f_σ are C^2 and such that $\det Df_\sigma(x) \ne 0$ for all $x \in U$. We write

$$\psi(\omega) := \log|Df_{\omega_1}(F\omega)|.$$

Observe that $\psi \le \log a < 0$. By making the set U a bit smaller, if necessary, we may and will assume that the f_σ extend to C^2-functions on $\bar{U}$ and that $\inf_{x\in U}|Df_\sigma(x)| > 0$.

A dynamical quantity that we are going to compare to the geometrically defined number $HD(F\Omega)$ is

$$t_0 := \sup\left\{\frac{h(\mu)}{\mu(-\psi)} : \mu \in \mathcal{M}(T)\right\}.$$

By its definition, $h(\mu) + t_0\mu(\psi) \le 0$ for all $\mu \in \mathcal{M}(T)$, whence $p(t_0\psi) \le 0$. Below we identify t_0 as the unique zero of the function $t \mapsto p(t\psi)$.

6.5.3 Proposition *Let $\mathcal{F}$ be a nondegenerate IFS. Then*

a) *The functions $x \mapsto \log |Df_\sigma(x)|$, $\sigma \in \Sigma$, are uniformly Lipschitz-continuous on U.*

b) *ψ is a regular local energy function on Ω,*

c) *for each $t \in \mathbb{R}$ there is a unique $\mu_t \in ES(t\psi, T)$,*

d) *the pressure function $t \mapsto p(t\psi)$ is differentiable with derivative $\frac{d}{dt}p(t\psi) = \mu_t(\psi) \leq \log a < 0$, and*

e) *$t_0 \in (0, \infty)$ and t_0 is the unique zero of $t \mapsto p(t\psi)$.*

Proof: If $\mathcal{F}$ is nondegenerate, then $x \mapsto |Df_\sigma(x)|$ is a strictly positive C^1-function on $\bar{U}$ for each $\sigma \in \Sigma$. Hence ψ is Hölder-continuous by Lemma 6.4.2, so it is in particular a regular local energy function. Therefore there exists for each $t \in \mathbb{R}$ a unique equilibrium state μ_t for $t\psi$ which is at the same time the unique Gibbs measure for $t\psi$, see Corollary 5.3.2. By Theorems 4.3.3 and 4.3.5,

$$\frac{d}{dt}p(t\psi) = \lim_{s \to 0} \frac{1}{s}(p((t+s)\psi) - p(t\psi)) = \mu_t(\psi)$$

where $\mu_t(\psi) \leq \sup \psi \leq \log a < 0$. As $p(0) = \log |\Sigma| > 0$ by Example 4.2.6, the existence of a unique positive zero t_1 for $t \mapsto p(t\psi)$ follows from the intermediate value theorem. $t_1 \leq t_0$ as $p(t_0\psi) \leq 0$, and $t_0 \leq t_1$ as $p(t_1\psi) = 0$ implies $h(\mu) + t_1\mu(\psi) \leq 0$ for all $\mu \in \mathcal{M}(T)$. □

In order to compare the number t_0 with $HD(F\Omega)$ we make a strong geometric conformality assumption:

6.5.4 Definition *The IFS $\mathcal{F}$ is conformal, if it is nondegenerate, if all f_σ are injective, and if $|(Df_\sigma(x))v| = |Df_\sigma(x)| \cdot |v|$ for all $x \in U$, $v \in \mathbb{R}^d$ and $\sigma \in \Sigma$.*

6.5.5 Remark How restrictive the conformality assumption is depends heavily on the dimension. In $d = 1$ it just means that each f_σ is C^2 without critical point. In $d = 2$, conformal maps are either holomorphic or antiholomorphic. In $d \geq 3$, however, a theorem of Liouville asserts that each conformal mapping is a *Möbius transformation*, i.e., a composition of finitely many reflections in spheres or hyperplanes. (This includes all translations, stretchings and orthogonal transformations.) In any case $Df_\sigma(x)/|Df_\sigma(x)|$ is a unitary matrix and

$$|\det Df_\sigma(x)| = |Df_\sigma(x)|^d$$

for all σ and x if $\mathcal{F}$ is conformal. In particular, $|D(f_\tau \circ f_\sigma)(x)| = |Df_\tau(f_\sigma x)| \times |Df_\sigma(x)|$ for all $\sigma, \tau \in \Sigma$. Possible starting points for further reading on conformal maps are [59, Section I.1] and [58, Remark 5.8]. A good review of Hausdorff dimension theory for nonconformal IFS is [21]. ◇

Below we prove

6.5.6 Proposition *Let $\mathcal{F}$ be a conformal IFS. Then*

$$HD(F\mu_{t_0}) \leq \sup\{HD(F\mu) : \mu \in \mathcal{M}(T)\} \leq HD(F\Omega) \leq t_0 \,.$$

Here t_0 is the zero of $p(t\psi)$ from Proposition 6.5.3.

In order to close this chain of inequalities, we need one further assumption which implies that there is essentially a one-to-one correspondence between sequences $\omega_{-\infty}^0$ and points $F(\omega_{-\infty}^0) \in F\Omega$. Denote by m Lebesgue measure on $\mathbb{R}^d$.

6.5.7 Definition *The IFS $\mathcal{F}$ satisfies the disjointness condition, if the compact invariant set K entering the definition of an IFS (Definition 6.4.1) is such that $m(K) > 0$ and $m(f_\sigma(K) \cap f_\tau(K)) = 0$ for all $\sigma, \tau \in \Sigma$.*

6.5.8 Proposition *Let $\mathcal{F}$ be a conformal IFS satisfying the disjointness condition. Then $t_0 \leq HD(F\mu_{t_0})$.*

These results are collected in

6.5.9 Theorem (Measures of maximal dimension)
Let $\mathcal{F}$ be a conformal IFS satisfying the disjointness condition with local energy function ψ as in Definition 6.5.4. Then the pressure function $t \mapsto p(t\psi)$ $(t \in \mathbb{R})$ has a unique zero at $t_0 = HD(F\Omega) > 0$ and for the unique equilibrium state $\mu_{t_0} \in ES(t_0\psi, T)$ we have

$$\lim_{r\to 0} \frac{\log\left((F\mu_{t_0})(B_r(x))\right)}{\log r} = HD(F\Omega) \quad \text{for all } x \in F\Omega. \tag{6.18}$$

(Here $B_r(x) = \{y \in \mathbb{R}^d : |x - y| < r\}$.) The measure μ_{t_0} is distinguished by the property that it is the unique ergodic measure $\mu \in \mathcal{M}(T)$ that maximizes the dimension $HD(F\mu)$, i.e., it is the only ergodic measure for which $HD(F\mu) = HD(F\Omega)$.

The theorem follows from the preceding propositions except for the identity (6.18). Before we proceed to the proofs, we illustrate the theorem by the following example:

6.5.10 Example (Sierpinski gasket) Recall the IFS from Example 6.4.4 where $\Sigma = \{0, 1, 2\}$, K is the equilateral triangle with corners $A_\sigma = e^{2\pi i \frac{\sigma}{3}}$ in the complex plane and f_σ is the linear contraction with centre A_σ and contraction rate a_σ ($\sigma \in \Sigma$). Then $\psi(\omega) = \log a_{\omega_1}$, and the unique equilibrium state $\mu_t \in ES(t\psi, T)$ is the Bernoulli measure with one-dimensional marginal probabilities $\tilde{q}_\sigma = a_\sigma^t\, e^{-p(t\psi)}$, $p(t\psi) = \log \sum_{\sigma\in\Sigma} a_\sigma^t$, see Example 4.2.2. This IFS is clearly conformal, and if

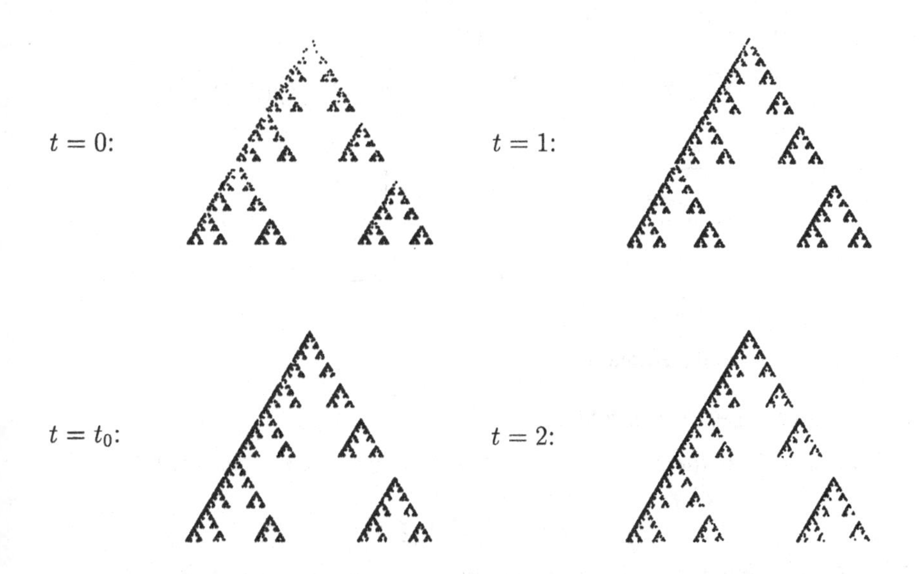

Figure 6.2: Random trajectories of the IFS from Example 6.5.10 that are generated using measures μ_t at four different values of t. Here $t_0 = HD(F\Omega) \approx 1.335$.

we assume that the sum of any two of the a_σ is less than or equal to 1, then the disjointness condition is trivially satisfied.

Therefore $HD(F\Omega)$ is the unique t for which $\sum_{\sigma\in\Sigma} a_\sigma^t = 1$. This t is also known as the *self-similarity exponent* in the literature. In Figure 6.2 we plotted random trajectories of length 3000 for the system with $a_0 = 0.6$, $a_1 = 0.4$ and $a_2 = 0.3$. The four plots display trajectories under the laws μ_t with $t = 0$, i.e., equal probabilities, $t = 1$, i.e., probabilities proportional to a_σ, $t = HD(F\Omega)$, and $t = d$, i.e., probabilities proportional to the Euclidean volume contraction. It is evident that $t = HD(F\Omega)$ yields the most uniform distribution of the trajectory. $\diamond$

For the proofs of the propositions and of the theorem we need the following two lemmas.

6.5.11 Lemma *Suppose that $\mathcal{F}$ is a conformal IFS. Let L be a common Lipschitz constant for the functions $x \mapsto \log|Df_\sigma(x)|$ on U. Then*

$$\sup \frac{1}{|x-y|} \left|\log|Df(\omega^0_{-n}, x)| - \log|Df(\omega^0_{-n}, y)|\right| \le \frac{L}{1-a}$$

where the supremum extends over all $\omega \in \Omega$, $0 < n \in \mathbb{Z}$ and $x, y \in U$.

Proof:

$$
\begin{aligned}
&\left|\log|Df(\omega_{-n}^0, x)| - \log|Df(\omega_{-n}^0, y)|\right| \\
&\le \sum_{k=-n+1}^{0} \left|\log|Df_{\omega_k}(f(\omega_{-n}^{k-1}, x))| - \log|Df_{\omega_k}(f(\omega_{-n}^{k-1}, y))|\right| \\
&\le L \sum_{k=-n+1}^{0} \left|f(\omega_{-n}^{k-1}, x) - f(\omega_{-n}^{k-1}, y)\right| \\
&\le L \sum_{k=-n+1}^{0} a^{k-1+n}\,|x-y| \le \frac{L}{1-a}\,|x-y|
\end{aligned}
$$

by Remark 6.5.5 and Lemma 6.4.2. □

6.5.12 Lemma *Let $\mathcal{F}$ be a conformal IFS.*

1. *For each $t > t_0$ there is $r(t) > 0$ such that for $0 < r < r(t)$ and for all $x \in \mathbb{R}^d$ with $B_r(x) \cap F\Omega \neq \emptyset$*

$$(F\mu_{t_0})(B_{2r}(x)) \ge r^t \,. \tag{6.19}$$

2. *If $\mathcal{F}$ also satisfies the disjointness condition, then for each $t < t_0$ there is $r(t) > 0$ such that for $0 < r < r(t)$ and for all $x \in \mathbb{R}^d$*

$$(F\mu_{t_0})(B_r(x)) \le r^t \,. \tag{6.20}$$

Proof: For $\omega \in \Omega$ denote $[\omega_{-n}^0] := [\omega_k]_{k\in\{-n+1,\dots,0\}}$. Fix $x \in \mathbb{R}^d$ and $r > 0$ such that $B_r(x) \cap F\Omega \neq \emptyset$. For $\omega \in F^{-1}(B_r(x))$ let

$$n(\omega) := \min\{n > 0 : f(\omega_{-n}^0, K) \subseteq B_{2r}(x)\} \,.$$

Observe that for all $n \ge 1$ and $\omega \in F^{-1}(B_r(x))$

- ▷ $F(\omega) = f(\omega_{-n}^0, F(T^{-n}\omega)) \in f(\omega_{-n}^0, K)$ so that $f(\omega_{-n}^0, K) \cap B_r(x) \neq \emptyset$,
- ▷ $\mathrm{diam}(f(\omega_{-n}^0, K)) \le a^n \cdot \mathrm{diam}(K)$ by Lemma 6.4.2,
- ▷ $\mathrm{diam}(f(\omega_{-n}^0, K)) \ge b^n \cdot \mathrm{diam}(K)$, where $b := \inf\{|f_\sigma(x) - f_\sigma(y)|/|x-y| : x \neq y \in K, \sigma \in \Sigma\} > 0$ since $\mathcal{F}$ is conformal.

The first two statements imply that $n(\omega)$ is bounded for $\omega \in F^{-1}(B_r(x))$. In particular, $V := \{\omega_{-n(\omega)}^0 : F(\omega) \in B_r(x)\}$ is a finite set of words over the alphabet Σ. It follows also that

$$b\,r \le \mathrm{diam}(f(\omega_{-n(\omega)}^0, K)) \le a^{n(\omega)} \cdot \mathrm{diam}(K) \tag{6.21}$$

and

$$b^{n(\omega)} \cdot \mathrm{diam}(K) \le \mathrm{diam}(f(\omega_{-n(\omega)}^0, K)) \le 4r \,. \tag{6.22}$$

We claim that $[\omega^0_{-n(\omega)}] \cap [\tilde{\omega}^0_{-n(\tilde{\omega})}] = \emptyset$ if $\omega^0_{-n(\omega)}, \tilde{\omega}^0_{-n(\tilde{\omega})} \in V$ and $\omega^0_{-n(\omega)} \neq \tilde{\omega}^0_{-n(\tilde{\omega})}$. Suppose for a contradiction that $[\omega^0_{-n(\omega)}] \cap [\tilde{\omega}^0_{-n(\tilde{\omega})}] \neq \emptyset$ and that w.l.o.g. $n(\omega) \leq n(\tilde{\omega})$. Then $\tilde{\omega}^0_{-n(\omega)} = \omega^0_{-n(\omega)}$ so that $f(\tilde{\omega}^0_{-n(\omega)}, K) = f(\omega^0_{-n(\omega)}, K) \subseteq B_{2r}(x)$ and hence $n(\tilde{\omega}) = n(\omega)$, i.e., $\omega^0_{-n(\omega)} = \tilde{\omega}^0_{-n(\tilde{\omega})}$. For the sake of notational convenience we abbreviate $\omega^0_{-n(\omega)}$ as ω^*.

Observe that by construction of the set V

$$F\Omega \cap B_r(x) \subseteq \bigcup_{v \in V} f(v, K) \subseteq B_{2r}(x) .$$

Fix $\omega^* \in V$. In view of the distortion estimate from Lemma 6.5.11,

$$\begin{aligned}
\operatorname{diam}(f(\omega^*, K)) &= \sup_{y,z \in K} \left| \int_0^1 Df(\omega^*, (1-t)y + tz)\, dt \right| \cdot |z - y| \\
&\leq |Df(\omega^*, x)| \cdot \exp \frac{L \cdot \operatorname{diam}(U)}{1-a} \cdot \operatorname{diam}(K) \\
&= e^{\psi_{n(\omega)}(T^{-n(\omega)}\omega)} \cdot M \cdot \operatorname{diam}(K)
\end{aligned} \tag{6.23}$$

where $M := \exp \frac{L \cdot \operatorname{diam}(U)}{1-a}$, so that in view of (6.21) and Theorem 5.2.4, which gives a precise approximation for the μ_{t_0}-measure of cylinder sets,

$$\begin{aligned}
& a^{-(t-t_0)\,n(\omega)} \operatorname{diam}(f(\omega^*, K))^t \\
&= \operatorname{diam}(f(\omega^*, K))^{t_0} \cdot \left(a^{-n(\omega)} \operatorname{diam}(f(\omega^*, K))\right)^{t-t_0} \\
&\leq e^{t_0\, \psi_{n(\omega)}(T^{-n(\omega)}\omega)} \cdot M^{t_0} \cdot \operatorname{diam}(K)^{t_0} \cdot \operatorname{diam}(K)^{t-t_0} \\
&\leq M^{t_0}\, e^{C_{t_0\psi}}\, e^{P_{n(\omega),1}(t_0\psi)}\, \mu_{t_0}[\omega^*] \operatorname{diam}(K)^t \quad \text{for } t > t_0.
\end{aligned} \tag{6.24}$$

Next,

$$\begin{aligned}
m(f(\omega^*, K)) &= \int_K |Df(\omega^*, y)|^d\, dm(y) \qquad \text{(see Remark 6.5.5)} \\
&\geq |Df(\omega^*, x)|^d \cdot M^{-d} \cdot m(K) \\
&= e^{d \cdot \psi_{n(\omega)}(T^{-n(\omega)}\omega)} \cdot M^{-d} \cdot m(K) ,
\end{aligned} \tag{6.25}$$

and as $\operatorname{diam}(f(\omega^*, K)) \geq m(f(\omega^*, K))^{1/d}$, it follows that

$$\begin{aligned}
& a^{-(t-t_0)\,n(\omega)} \operatorname{diam}(f(\omega^*, K))^t \\
&= \operatorname{diam}(f(\omega^*, K))^{t_0} \cdot \left(a^{-n(\omega)} \operatorname{diam}(f(\omega^*, K))\right)^{t-t_0} \qquad (6.26) \\
&\geq M^{-t_0}\, e^{-C_{t_0\psi}}\, e^{P_{n(\omega),1}(t_0\psi)}\, \mu_{t_0}[\omega^*]\, m(K)^{t_0/d} \operatorname{diam}(K)^{t-t_0} \quad \text{for } t < t_0.
\end{aligned}$$

(In fact, for these estimates we have to apply Theorem 5.2.4 in a slightly modified version, because here we are dealing with "one-sided" cylinder sets whereas that theorem makes statements about "two-sided" cylinder sets. Because of the stationarity of μ_t this makes no essential difference.)

Suppose now that $t > t_0$. As $\lim_{n\to\infty} \frac{1}{n} P_{n,1}(t_0\psi) = p(t_0\psi) = 0$ and $a^{t-t_0} < 1$, we have

$$\lim_{n\to\infty} a^{(t-t_0)n} M^{t_0} e^{C_{t_0\psi}} e^{P_{n,1}(t_0\psi)} \operatorname{diam}(K)^t = 0 .$$

Therefore, observing that $n(\omega) \geq \text{const} \cdot |\log r|$ by (6.22), it follows from (6.21) and (6.24) that there is some $r(t) > 0$ such that for $r < r(t)$

$$\begin{aligned} r^t &\leq b^{-t} (\operatorname{diam}(f(\omega^*, K)))^t \leq \mu_{t_0}[\omega^*] \leq (F\mu_{t_0})(F[\omega^*]) \\ &\leq (F\mu_{t_0})(f(\omega^*, K)) \leq (F\mu_{t_0})(B_{2r}(x)) . \end{aligned}$$

This is inequality (6.19).

The assertion for $t < t_0$ is a bit more difficult to prove, because now we must cover $B_r(x)$ by sets $f(\omega^*, K)$, $\omega^* \in V$, thereby controlling the cardinality of V uniformly in x and r. As the assertion is trivially true if $B_r(x) \cap F\Omega = \emptyset$, we can assume that this intersection is nonempty.

A simple inductive argument shows that the disjointness condition implies $m(f(v, K) \cap f(v', K)) = 0$ for all $v, v' \in V$, $v \neq v'$. As $\bigcup_{v\in V} f(v, K) \subseteq B_{2r}(x)$, it follows that $\sum_{v\in V} m(f(v, K)) \leq \text{const}_d \cdot r^d$. On the other hand, if $v = \omega^* \in V$, then it follows from (6.25), (6.23) and (6.21) that

$$m(f(v, K)) \geq M^{-2d} \frac{m(K)}{\operatorname{diam}(K)^d} b^d r^d .$$

Therefore $\operatorname{card}(V) \leq C := \text{const}_d M^{2d} b^{-d} \cdot \frac{\operatorname{diam}(K)^d}{m(K)}$ independently of x and r.

Now we can proceed as before: As $\lim_{n\to\infty} \frac{1}{n} P_{n,1}(t_0\psi) = p(t_0\psi) = 0$, as $a^{t-t_0} > 1$ and as $n(\omega) \geq \text{const} \cdot |\log r|$ by (6.22), it follows from (6.22) and (6.26) that there is some $r(t) > 0$ such that if $r < r(t)$ then

$$\begin{aligned} r^t &\geq \frac{1}{4^t C} \sum_{\omega^*\in V} (\operatorname{diam}(f(\omega^*, K)))^t \geq \sum_{\omega^*\in V} \mu_{t_0}[\omega^*] \\ &\geq \mu_{t_0}(F^{-1} B_r(x)) = (F\mu_{t_0})(B_r(x)) , \end{aligned}$$

i.e., inequality (6.20). □

Proof of Proposition 6.5.6: Consider $t > t_0$, fix $r < r(t)$ and let $\{x_1, \ldots, x_N\}$ be a maximal r-separated subset of $F\Omega$. (Here r-separated means $(1, r)$-separated in the sense of Definition 4.4.1.) Then the balls $U_n := B_r(x_n)$, $n = 1, \ldots, N$, cover $F\Omega$, they all have nonempty intersection with $F\Omega$, and the balls $B_{\frac{r}{2}}(x_n)$ are pairwise disjoint. Therefore

$$\sum_{n=1}^{N} (\operatorname{diam}(U_n))^t = 8^t \sum_{n=1}^{N} \left(\frac{r}{4}\right)^t \leq 8^t \sum_{n=1}^{N} (F\mu_{t_0})(B_{\frac{r}{2}}(x_n)) \leq 8^t\, (F\mu_{t_0})(\mathbb{R}^d) = 8^t .$$

Hence $\mathcal{H}^t_\epsilon(F\Omega) \leq 8^t$ for all $\epsilon < r(t)$ so that $t \geq HD(F\Omega)$. Since this holds for all $t > t_0$, it follows that $HD(F\Omega) \leq t_0$. □

Proof of Proposition 6.5.8: Consider $t > t_0$ and let $A \subseteq F\Omega$ be any measurable set with $(F\mu_{t_0})(A) = 1$. Let $(U_n)_n$ be a covering of A by balls $U_n = B_{r_n}(x_n)$ with radii $r_n < r(t)$. Then

$$\sum_{n>0}(\mathrm{diam}(U_n))^t = 2^t \sum_{n>0}(r_n)^t \geq 2^t \sum_{n>0}(F\mu_{t_0})(B_{r_n}(x_n)) \geq 2^t\,(F\mu_{t_0})(A) = 2^t\,.$$

Hence $\mathcal{H}^t_\epsilon(A) \geq 2^t > 0$ for all $\epsilon < r(t)$ so that $t \leq HD(A)$. As A was any measurable set with $(F\mu_{t_0})(A) = 1$ we conclude that $t \leq HD(F\mu_{t_0})$. Since this holds for all $t < t_0$, it follows that $t_0 \leq HD(F\mu_{t_0})$. □

Proof of Theorem 6.5.9: The assertions about the pressure function and its zero and the identity $HD(F\mu_{t_0}) = HD(F\Omega)$ follow from the foregoing propositions. Equation (6.18) is an immediate consequence of Lemma 6.5.12.

It remains to prove that μ_{t_0} is the unique ergodic measure in $\mathcal{M}(T)$ that maximizes $HD(F\mu)$. So suppose that $HD(F\mu) = HD(F\Omega) = t_0$ for some ergodic $\mu \in \mathcal{M}(T)$. As $ES(t_0\psi, T) = \{\mu_{t_0}\}$ by Proposition 6.5.3, it suffices to show that

$$h(\mu) + t_0\mu(\psi) \geq 0\,. \tag{6.27}$$

Fix a measurable set $A \subseteq F\Omega$ with $(F\mu)(A) = 1$. Let $\delta > 0$ and $j \in \mathbb{N}$. By Birkhoff's ergodic theorem and by the Shannon–McMillan–Breiman theorem there are a measurable $B_j \subseteq F^{-1}A$ with $\mu(B_j) > 1-2^{-j}$ and an integer $n(j,\delta) > 0$ such that for all $n > n(j,\delta)$ and all $\omega \in B_j$

$$\Bigl|\underbrace{\frac{1}{n}\sum_{k=1}^{n}\psi(T^{-k}\omega)}_{\psi_n(T^{-n}\omega)} - \mu(\psi)\Bigr| < \delta \quad\text{and}\quad \Bigl|-\frac{1}{n}\log\mu[\omega^0_{-n}] - h(\mu)\Bigr| < \delta\,.$$

Let $n > n(j,\delta)$ and fix a finite subset V_n of B_j such that the cylinder sets $[\omega^0_{-n}]$, $\omega \in V_n$, are pairwise disjoint and $B_j \subseteq \bigcup_{\omega\in V_n}[\omega^0_{-n}]$. Let $r(\omega) := \mathrm{diam}(F[\omega^0_{-n}])$. Then $r(\omega) \leq \epsilon_n := a^n \cdot \mathrm{diam}(F\Omega)$ by Lemma 6.4.2 and

$$F(B_j) \subseteq \bigcup_{\omega\in V_n} F[\omega^0_{-n}] \subseteq \bigcup_{\omega\in V_n} B_{r(\omega)}(F\omega)\,.$$

In view of the distortion estimates from Lemma 6.5.11,

$$\begin{aligned} r(\omega) &\leq \text{const}\cdot|Df(\omega^0_{-n}, F(T^{-n}\omega))| \\ &= \text{const}\cdot \prod_{k=-n+1}^{0} |Df_{\omega_k}(f(\omega^{k-1}_{-n}, F(T^{-n}\omega)))| \end{aligned}$$

$$= \text{const} \cdot \prod_{k=-n+1}^{0} |Df_{\omega_k}(F(T^{k-1}\omega))|$$
$$= \text{const} \cdot e^{\psi_n(T^{-n}\omega)} \,.$$

Hence

$$\begin{aligned}
\mathcal{H}^t_{\epsilon_n}(FB_j) &\leq \sum_{\omega \in V_n} \left(\text{diam}(B_{r(\omega)}(F\omega))\right)^t = 2^t \sum_{\omega \in V_n} r(\omega)^t \\
&\leq \text{const} \cdot \sum_{\omega \in V_n} e^{t\,\psi_n(T^{-n}\omega)} \qquad (6.28)\\
&\leq \text{const} \cdot \sum_{\omega \in V_n} e^{n(t\mu(\psi)+t\delta)}\, e^{n(h(\mu)+\delta)}\, \mu[\omega^0_{-n}] \\
&\leq \text{const} \cdot e^{n(t\mu(\psi)+h(\mu)+(t+1)\delta)} \qquad (6.29)
\end{aligned}$$

as the $[\omega^0_{-n}]$ figuring in the sum are pairwise disjoint.

Consider now $t < t_0$. As $HD(F\mu) = t_0$ and as $(F\mu)(\bigcup_j FB_j) \geq \sup_j \mu(B_j) = 1$, also $HD(\bigcup_j FB_j) \geq t_0$. It follows from Lemma 6.5.2 that $\sup_j HD(FB_j) \geq t_0$ so that

$$\sup_n \mathcal{H}^t_{\epsilon_n}(FB_j) = \mathcal{H}^t(FB_j) = \infty \quad \text{for some } j.$$

In view of (6.29) this is possible only if $t\mu(\psi) + h(\mu) + (t+1)\delta > 0$, and as $\delta > 0$ was arbitrary this implies $t\mu(\psi) + h(\mu) \geq 0$. As $t < t_0$ was arbitrary this proves the claim (6.27). □

Appendix

In this appendix we collect a number of basic facts from analysis, measure theory and probability theory.

A.1 Lipschitz-continuous functions

A.1.1 Let (X_1, ρ_1), (X_2, ρ_2) be metric spaces. A function $f : X_1 \to X_2$ is *Lipschitz-continuous* with Lipschitz constant $L > 0$, if $\rho_2(f(x), f(y)) \leq L \cdot \rho_1(x, y)$ for all $x, y \in X_1$. More generally, f is *Hölder-continuous*, if there are constants $a > 0$ and $H > 0$ such that $\rho_2(f(x), f(y)) \leq H \cdot \rho_1(x, y)^a$ for all $x, y \in X_1$.

A.1.2 Let (X, ρ) be a compact metric space and denote by $C(X)$ and *Lip*(X) the sets of all continuous and all Lipschitz-continuous functions $f : X \to \mathbb{R}$, respectively. Then *Lip*(X) is a dense subset of $C(X)$ with respect to the sup-norm on $C(X)$. An elementary proof of this fact is a good exercise in metric spaces, but the denseness follows also from the Stone–Weierstrass approximation theorem [16, Theorem IV.6.16].

A.1.3 The denseness of *Lip*(X) in $C(X)$ is complemented by the *compactness* of the set $\{f \in \textit{Lip}(X) : f \text{ has Lipschitz constant } L\}$ in $C(X)$ for any $L > 0$. As this set is obviously closed in $C(X)$, its compactness is a special case of the theorem of *Arzelà and Ascoli*, which asserts that each set of uniformly bounded and uniformly continuous real functions on X is relatively compact in $C(X)$ [16, Theorem IV.6.7].

A.2 Some convex analysis in $\mathbb{R}^d$

A.2.1 Let V be a linear space. A set $C \subseteq V$ is *convex*, if $\alpha u + (1 - \alpha)v \in C$ whenever $u, v \in C$ and $0 < \alpha < 1$.

Let $I = [-\infty, +\infty)$ or $I = (-\infty, +\infty]$. A function $f : C \to I$ is *convex*, if $f(\alpha u + (1 - \alpha)v) \leq \alpha f(u) + (1 - \alpha)f(v)$ for all $u, v \in C$ and $0 < \alpha < 1$. It is strictly convex if equality holds only for $u = v$.

A function $f : C \to I$ is *concave*, if $-f$ is convex, and $f : C \to \mathbb{R}$ is *convex affine*, if it is both convex and concave.

If $V = \mathbb{R}$ and if $f : C \to \mathbb{R}$ is of class C^2, then f is (strictly) convex if and only if $f'' \geq 0$ ($f'' > 0$).

A.2.2 (Elementary Jensen inequality) Let $f : C \to I$ be a convex function, let $x_1, \ldots, x_k \in C$, $0 < \alpha_1, \ldots, \alpha_k < 1$ and $\sum_{i=1}^k \alpha_i = 1$. It follows by induction that $f(\sum_{i=1}^k \alpha_i x_i) \leq \sum_{i=1}^k \alpha_i f(x_i)$. If f is strictly convex then equality holds if and only if $x_1 = \cdots = x_k$. (This is an elementary version of *Jensen's inequality* which is generalized below.)

A.2.3 If $C \subset \mathbb{R}^k$ is a compact convex set and if $z \in \mathbb{R}^k \setminus C$, then there is a hyperplane that separates z from C. Formally: there are real numbers $a_1, \ldots, a_k$ such that $\sum_{i=1}^k a_i x_i < \sum_{i=1}^k a_i z_i$ for all $x \in C$.

A.3 Non-negative matrices

A non-negative square matrix A is said to be *primitive*, if there exists a positive integer k such that all entries of the matrix A^k are strictly positive. The following theorem, known as the theorem of Frobenius and Perron, is part of [56, Theorem 1.1]

A.3.1 Suppose A is a non-negative primitive matrix. Then there exists an eigenvalue $r > 0$ such that

1. $r > |\lambda|$ for any eigenvalue $\lambda \neq r$,
2. the left and right eigenvectors u^t and v associated with r are strictly positive and unique up to constant multiples,
3. r is a simple root of the characteristic equation of A.

A.3.2 Let A, r, u^t, and v be as in the theorem. We may assume without loss of generality that $\sum_i u_i = 1$ and $\sum_i u_i v_i = 1$. Then $(vu^t)^2 = vu^t$, i.e., vu^t is a projection matrix with left and right eigenvectors u^t and v. Therefore the sequence of matrices $(r^{-k}A^k)_{k>0}$ converges elementwise to the matrix vu^t.

Define a positive matrix Q by $q_{ij} = a_{ij}\,\frac{v_j}{rv_i}$, i.e., $Q = r^{-1}\operatorname{diag}(v)^{-1}A\operatorname{diag}(v)$. Then $\sum_j q_{ij} = 1$ for all i, i.e., Q is a stochastic matrix. Iterates of A and Q are closely related: $Q^k = r^{-k}\operatorname{diag}(v)^{-1}A^k\operatorname{diag}(v)$. In particular Q^k converges elementwise to the matrix whose rows are all identical to the probability vector $(u_i v_i)_i$.

A.4 Some facts from probability and integration

Most of the definitions and results given below can be found (with slight variations) in any introductory text to measure and probability theory, e.g., [2], [5]. Only for some particular results do we give precise references.

σ-algebras and their generators

A.4.1 Let X be a set. A collection $\mathcal{A}$ of subsets of X is a *σ-algebra*, if i) $X \in \mathcal{A}$, ii) $X \setminus A \in \mathcal{A}$ whenever $A \in \mathcal{A}$, and iii) $\bigcup_{n\geq 1} A_n \in \mathcal{A}$ for each sequence of sets $A_n \in \mathcal{A}$. The pair $(X, \mathcal{A})$ is called a *measurable space* and elements of $\mathcal{A}$ are called *measurable sets*.

If property iii) is required only for pairwise disjoint sets A_n, then $\mathcal{A}$ is called a *Dynkin system*.

A.4.2 Let $\mathcal{E}$ be any collection of subsets of X. By $\sigma(\mathcal{E})$ and $\delta(\mathcal{E})$ we denote the smallest σ-algebra and the smallest Dynkin system containing $\mathcal{E}$, respectively. We also say these are the σ-algebra and the Dynkin system generated by $\mathcal{E}$. If $\mathcal{E}$ is $\cap$-*stable*, i.e., if $A \cap B \in \mathcal{E}$ whenever $A, B \in \mathcal{E}$, then $\sigma(\mathcal{E}) = \delta(\mathcal{E})$ (see [2, Theorem 2.4] or [5, Theorem 3.2]).

A.4.3 Let (X, ρ) be a metric space. The *Borel σ-algebra* $\mathcal{B}$ of X is the σ-algebra generated by all open subsets of X. If (X, ρ) is separable, then there is a countable $\cap$-stable collection $\mathcal{E}$ of subsets of X that generates $\mathcal{B}$. The easy proof of this fact can be found, e.g., in [25, Lemma 3.2.1].

A.4.4 Examples: a) If $(X, \mathcal{B})$ is a nontrivial subinterval of $[-\infty, +\infty]$ endowed with its Borel σ-algebra, then the set $\mathcal{E}$ of all intervals of the form $X \cap [-\infty, a)$ with rational $a \in X$ is an $\cap$-stable generator of $\mathcal{B}$, because each open subinterval of X belongs to $\sigma(\mathcal{E})$.
b) Let Σ be a finite set, $G = \mathbb{Z}^d$ or $G = \mathbb{Z}_+^d$ and $\Omega = \Sigma^G$ as in Example 2.2.10. Endowed with the product topology of the discrete topology on Σ the space Ω is a compact topological space (Tychonov's theorem). This topology is also generated by the metric $d(\omega, \omega') := 2^{-n(\omega,\omega')}$ where $n(\omega, \omega') := \sup\{n \in \mathbb{N} : \omega_g = \omega'_g \text{ for } g \in \Lambda_n\}$ and $\Lambda_n := \{g \in G : |g_1|, \ldots, |g_d| < n\}$ is a "cube" of side length n (if $G = \mathbb{Z}_+^d$) or $2n - 1$ (if $G = \mathbb{Z}^d$) (see [16, Section I.8]). Let $\mathcal{C}_n$ be the family of *cylinder sets* that depend only on coordinates $g \in \Lambda_n$, i.e.,

$$\mathcal{C}_n := \{[\sigma_g]_{g\in\Lambda_n} : \sigma_g \in \Sigma \; (g \in \Lambda_n)\}$$

where $[\sigma_g]_{g\in\Lambda_n} := \{\omega \in \Omega : \omega_g = \sigma_g \; (g \in \Lambda_n)\}$, and set $\mathcal{C} := \bigcup_n \mathcal{C}_n$. $\mathcal{C}$ is a countable $\cap$-stable family generating the Borel σ-algebra $\mathcal{B}$ of Ω.

Finite measures

A.4.5 Let $(X, \mathcal{A})$ be a measurable space. A function $\mu : \mathcal{A} \to [0, +\infty)$ is a *finite measure*, if i) $\mu(\emptyset) = 0$ and ii) $\mu(\bigcup_{n=1}^{\infty} A_n) = \sum_{n=1}^{\infty} \mu(A_n)$ for each sequence of pairwise disjoint sets $A_n \in \mathcal{A}$. The triple $(X, \mathcal{A}, \mu)$ is called a *finite measure space.* If $\mu(X) = 1$, then μ is a *probability measure.* In this case $(X, \mathcal{A}, \mu)$ is called a *probability space.*

Warning! Henceforth we skip the adjective "finite", because we will not mention infinite measures. Many of the further results are not true for infinite measures!

A.4.6 Let $(X, \mathcal{A}, \mu)$ be a measure space. A set $A \in \mathcal{A}$ is an *atom,* if it contains no subset $A' \in \mathcal{A}$ with $0 < \mu(A') < \mu(A)$. The measure μ is *purely atomic,* if there are at most countably many atoms A_k such that $\mu(X \setminus \bigcup_k A_k) = 0$. It is *non-atomic* if it has no atoms.

A.4.7 Let $(X, \mathcal{A})$ be a measurable space. Suppose that $\mathcal{A}$ is generated by the $\cap$-stable collection $\mathcal{E}$. If μ_1, μ_2 are two measures on $(X, \mathcal{A})$ such that $\mu_1(E) = \mu_2(E)$ for all $E \in \mathcal{E}$, then $\mu_1 = \mu_2$ (see [2, Theorem 5.4] or [5, Theorem 3.3]).

We do not discuss here how measures with specific properties are actually constructed (i.e., the existence problem). But we mention two examples:

A.4.8 Examples:
a) With each finite hyper-rectangle $A \subseteq \mathbb{R}^k$ we can associate its k-dimensional volume vol(A). Let X be a fixed hyper-rectangle. Then the set of all hyper-rectangles contained in X is an $\cap$-stable generator of the Borel subsets $\mathcal{B}$ of X, and there is a probability measure m on X that assigns the value $m(A) = \frac{\text{vol}(A)}{\text{vol}(X)}$ to each hyper-rectangle $A \subseteq X$. It is the *normalized Lebesgue measure* on X.
b) On $(\Omega = \Sigma^G, \mathcal{B})$ from Example A.4.4-b one can define *product measures* μ in the following way: Let $(p_\sigma : \sigma \in \Sigma)$ be a probability vector on Σ. There exists a probability measure μ on Ω such that $\mu([\sigma_g]_{g \in \Lambda_n}) = \prod_{g \in \Lambda_n} p_{\sigma_g}$ for each cylinder set $[\sigma_g]_{g \in \Lambda_n} \in \mathcal{C}_n$. More generally: if there is a finitely additive function μ defined on all cylinder sets and such that $\mu(\emptyset) = 0$ and $\mu(\Omega) = 1$, then it extends in a unique way to a probability measure on $(\Omega, \mathcal{B})$.

A.4.9 Let $(X, \mathcal{A}, \mu)$ be a measure space. For most of our considerations sets of measure zero can be neglected. If a property P that can be attributed to points $x \in X$ holds for all $x \in X$ except possibly for points in a set of μ-measure zero, we say that P holds for *μ-almost every x* (μ-a.e. x) or *μ-almost surely* (μ-a.s.). If two sets $A, B \in \mathcal{A}$ coincide up to a set of measure zero (i.e., if $\mu(A \Delta B) = 0$), then we also write $A = B \bmod \mu$.

Measurable functions

A.4.10 Let $(X_1, \mathcal{A}_1)$ and $(X_2, \mathcal{A}_2)$ be a measurable spaces. A mapping $f : X_1 \to X_2$ is *measurable*, if $f^{-1}A \in \mathcal{A}_1$ for each $A \in \mathcal{A}_2$. Obviously, compositions of measurable maps between suitable measurable spaces are again measurable.

A.4.11 Let $(X, \mathcal{A})$ be a measurable space. A function $f : X \to [-\infty, +\infty]$ is *measurable*, if $f^{-1}B \in \mathcal{A}$ for each measurable set $B \subseteq [-\infty, +\infty]$. f is already measurable if the sets $\{x \in X : f(x) < t\}$ belong to $\mathcal{A}$ for all $t \in \mathbb{R}$.

A.4.12 Pointwise sums, differences, products, quotients, limits, suprema, infima, etc. of measurable functions are again measurable. In particular, if f is measurable, then so are $f^+ := \max\{f, 0\}$ and $f^- := (-f)^+$.

A.4.13 Let (X_1, ρ_2) and (X_2, ρ_2) be metric spaces and let $\mathcal{B}_1, \mathcal{B}_2$ be their Borel σ-algebras. If $f : X_1 \to X_2$ is continuous, then f is measurable.

A.4.14 Let $(X, \mathcal{A})$ be a measurable space. $f : X \to \mathbb{R}$ is an *elementary function*, if there are $A_1, \ldots, A_n \in \mathcal{A}$ and $\alpha_1, \ldots, \alpha_n \in \mathbb{R}$ such that $f = \sum_{i=1}^n \alpha_i 1_{A_i}$. Obviously each elementary function is measurable, and it is not hard to show that each non-negative measurable function f is the pointwise supremum of an increasing sequence of elementary functions.

Integrable functions

A.4.15 Let $(X, \mathcal{A}, \mu)$ be a measure space. Integrals (with respect to μ) of measurable functions are defined in three steps:
1) If $f = \sum_{i=1}^n \alpha_i 1_{A_i}$ is an elementary function, then $\int f\, d\mu := \sum_{i=1}^n \alpha_i \mu(A_i)$.
2) If $f \geq 0$ is measurable, then $\int f\, d\mu := \sup\{\int g\, d\mu : g \text{ elementary}, g \leq f\}$.
3) If f is measurable and if $\int f^+\, d\mu < \infty$ or $\int f^-\, d\mu < \infty$, then $\int f\, d\mu := \int f^+\, d\mu - \int f^-\, d\mu$.
A measurable function is *integrable*, if $\int |f|\, d\mu < \infty$. The set of integrable functions is denoted by L^1_μ.

A.4.16 Pointwise sums, differences, maxima and minima of finitely many integrable functions are integrable. The integral is a linear, isotone map from L^1_μ to $\mathbb{R}$.

A.4.17 (Jensen inequality) Let $I \subseteq \mathbb{R}$ be an open interval and assume that $\varphi : I \to \mathbb{R}$ is convex. If $f \in L^1_\mu$ takes its values in I, then the integral of $\varphi \circ f$ is well defined and $\varphi(\int f\, d\mu) \leq \int \varphi \circ f\, d\mu$ with equality if and only if f is constant μ-a.e., see [2, Theorem 3.9] or [5, (21.14)].

Convergence theorems

Let $(X, \mathcal{A}, \mu)$ be a measure space.

A.4.18 (Fatou's lemma) Let $(f_n)_n$ be a sequence of measurable functions, $f_n \geq g$ for some $g \in L^1_\mu$. Then $\int \liminf_n f_n \, d\mu \leq \liminf_n \int f_n \, d\mu$.

A.4.19 (Monotone convergence theorem) Let $(f_n)_n$ be a sequence of measurable functions, $f_n \geq g$ for some $g \in L^1_\mu$. Suppose that $f_1 \leq f_2 \leq \ldots$. Then $\int \sup_n f_n \, d\mu = \sup_n \int f_n \, d\mu$, and $\sup_n f_n \in L^1_\mu$ if and only if $\sup_n \int f_n \, d\mu < \infty$.

A.4.20 (Dominated convergence theorem) Let $(f_n)_n$ be a sequence of measurable functions, $|f_n| \leq g$ for some $g \in L^1_\mu$. Suppose that $\lim_{n\to\infty} f_n(x) =: f(x)$ exists for μ-a.e. x. Then $f \in L^1_\mu$ and $\lim_{n\to\infty} \int |f - f_n| \, d\mu = 0$.

A.4.21 A family $\mathcal{F}$ of measurable functions is *uniformly integrable*, if

$$\lim_{M\to\infty} \sup_{f\in\mathcal{F}} \int_{\{|f|>M\}} |f| \, d\mu = 0$$

or, equivalently,

$$\lim_{M\to\infty} \sup_{f\in\mathcal{F}} \int (|f| - M)^+ \, d\mu = 0 \, .$$

(The equivalence of both conditions is proved, e.g., in [20, Theorem 1.14.7].) This concept is useful, because the conclusion of the theorem on dominated convergence remains true if the assumption that the functions f_n are dominated by an integrable g is replaced by the weaker assumption that the family $(f_n : n > 0)$ is uniformly integrable.

Two consequences of the monotone convergence theorem are:

A.4.22 (Borel–Cantelli lemma) Let $(A_n)_n$ be a sequence of measurable sets and suppose that $\sum_{n>0} \mu(A_n) < \infty$. Then μ-a.e. $x \in X$ belongs to at most finitely many of the A_n.

A.4.23 Consider $0 \leq h \in L^1_\mu$ with $\int h \, d\mu = 1$. Then $\nu(A) := \int_A h \, d\mu$ defines a probability measure on the measurable space $(X, \mathcal{A})$. It is the measure with *density* h w.r.t. μ. We write $\nu = h\mu$. The density h is mod μ uniquely determined by the measure ν.

Absolute continuity and mutual singularity

Let $(X, \mathcal{A})$ be a measurable space and let μ and ν be two measures on $(X, \mathcal{A})$.

A.4.24 The measure ν is *absolutely continuous* w.r.t. the measure μ ($\nu \ll \mu$), if $\nu(A) = 0$ whenever $A \in \mathcal{A}$ and $\mu(A) = 0$. The measures μ and ν are *equivalent* ($\mu \approx \nu$), if $\mu \ll \nu$ and $\nu \ll \mu$.

A.4.25 (Radon–Nikodym theorem) The probability measure ν is absolutely continuous w.r.t. μ, if and only if $\nu = h\mu$ for some density h. The density is denoted by $\frac{d\nu}{d\mu}$ (*Radon–Nikodym derivative*). As above, it is only mod μ uniquely determined by ν.

A.4.26 The measures μ and ν are *mutually singular* ($\mu \perp \nu$), if there is $A \in \mathcal{A}$ such that $\mu(A) = \nu(X \setminus A) = 0$.

A.4.27 (Lebesgue decomposition) Let μ and ν be two measures on $(X, \mathcal{A})$. Then ν can be written uniquely as a sum $\nu_1 + \nu_2$ of measures such that $\nu_1 \ll \mu$ and $\nu_2 \perp \mu$.

Measures on compact metric spaces

Let (X, ρ) be a compact metric space equipped with its Borel σ-algebra $\mathcal{B}$. A measure on the measurable space $(X, \mathcal{B})$ is called a *Borel measure*.

A.4.28 (Regularity of Borel measures) Let μ be a Borel measure on X. For $B \in \mathcal{B}$ and $\epsilon > 0$ there exists a compact $K \subseteq B$ such that $\mu(B \setminus K) < \epsilon$.

A.4.29 (Weak convergence) Let μ, μ_n be Borel measures on X. The sequence $(\mu_n)_n$ *converges weakly* to μ, if $\lim_{n\to\infty} \int f_n \, d\mu = \int f \, d\mu$ for all $f \in C(X)$. Equivalently: $\lim_{n\to\infty} \mu_n(B) = \mu(B)$ for all $B \in \mathcal{B}$ with $\mu(\partial B) = 0$, see [2, Corollary 31.3].

A.4.30 (Weak compactness) The set $\mathcal{M}$ of Borel probability measures on X is weakly sequentially compact, see [2, Corollary 31.3 with Remark 1]

A.4.31 (Riesz representation theorem) Let $C^*(X)$ denote the space of all real-valued bounded linear functionals on $C(X)$. There is a one-to-one correspondence between elements $\varphi^* \in C^*(X)$ and pairs (μ^+, μ^-) of mutually singular Borel measures on X. This correspondence is established by $\varphi^*(f) = \int f \, d\mu^+ - \int f \, d\mu^-$ for all $f \in C(X)$. See [2, Corollary 29.13].

Conditional expectation

Let $(X, \mathcal{A}, \mu)$ be a probability space.

A.4.32 Let $\mathcal{C}$ be a sub-σ-algebra of $\mathcal{A}$. For each $f \in L^1_\mu$ there exists a $\mathcal{C}$-measurable function $f_\mathcal{C}$ which is $\bmod \mu$ uniquely determined by the following property:

$$\int_C f\, d\mu = \int_C f_\mathcal{C}\, d\mu \quad \text{for each } C \in \mathcal{C}\,. \tag{A.1}$$

The $(\bmod \mu)$-equivalence class of $f_\mathcal{C}$ is called the *conditional expectation* and denoted by $E_\mu[f \mid \mathcal{C}]$ or simply by $E[f \mid \mathcal{C}]$. Each $f_\mathcal{C}$ that is $\mathcal{C}$-measurable and satisfies (A.1) is a *version of the conditional expectation* $E[f \mid \mathcal{C}]$. Equalities involving conditional expectations are automatically understood to hold μ-a.e., because their ingredients are only defined μ-a.e. It is easily seen that $\int g{\cdot}f\, d\mu = \int g{\cdot}E[f \mid \mathcal{C}]\, d\mu$ for all bounded $\mathcal{C}$-measurable $g : X \to \mathbb{R}$.

A.4.33 The map $f \mapsto E_\mu[f \mid \mathcal{C}]$ is a linear map from L^1_μ to $(\bmod \mu)$-equivalence classes of L^1_μ. It satisfies $\int |E_\mu[f \mid \mathcal{C}]|\, d\mu \leq \int |f|\, d\mu$.

A.4.34 For $A \in \mathcal{A}$ we write $\mu(A \mid \mathcal{C}) := E[1_A \mid \mathcal{C}]$ and call this the *conditional probability* of A given $\mathcal{C}$. Let β be a finite or countable partition of X into $\mathcal{C}$-measurable sets. Then $\mu(A \mid \mathcal{C}) = \sum_{B\in\beta} \frac{\mu(A\cap B)}{\mu(B)} \cdot 1_B$. This justifies the nomenclature "conditional probability".

A.4.35 (Jensen inequality, conditional) Let $I \subseteq \mathbb{R}$ be an open interval and assume that $\varphi : I \to \mathbb{R}$ is convex. If $\mathcal{C}$ is a sub-σ-algebra of $\mathcal{A}$, if $f \in L^1_\mu$ takes its values in I and if $\varphi \circ f \in L^1_\mu$, then $\varphi(E[f \mid \mathcal{C}]) \leq E[\varphi \circ f \mid \mathcal{C}]$, see [2, Theorem 15.3] or [5, (34.7)].

A.4.36 (Dominated convergence, conditional) Let $(f_n)_n$ be a sequence of measurable functions, $|f_n| \leq g$ for some $g \in L^1_\mu$ and let $\mathcal{C}$ be a sub-σ-algebra of $\mathcal{A}$. Suppose that $\lim_{n\to\infty} f_n(x) =: f(x)$ exists for μ-a.e. x. Then $\lim_{n\to\infty} E[f_n \mid \mathcal{C}] = E[f \mid \mathcal{C}]$ μ-a.s. [3, Assertion (15.14)].

A.4.37 (Martingale convergence theorem) Let $\mathcal{C}_1 \subset \mathcal{C}_2 \subset \cdots$ be an increasing sequence of sub-σ-algebras of $\mathcal{A}$ and set $\mathcal{C} = \sigma(\bigcup_n \mathcal{C}_n)$. Let $f_1, f_2, f_3, \ldots$ be a sequence of μ-integrable functions such that f_n is $\mathcal{C}_n$-measurable and $E[f_{n+1} \mid \mathcal{C}_n] = f_n$ (i.e., $(f_n, \mathcal{C}_n)_{n>0}$ is a *martingale*). Suppose that $\sup_n E[|X_n|] < \infty$. Then $f := \lim_{n\to\infty} f_n$ exists, is $\mathcal{C}$-measurable and is finite μ-a.e. If the sequence $(f_n)_n$ is uniformly integrable, then f_n converges in L^1_μ to f and $E[f \mid \mathcal{C}_n] = f_n$ for all n. See [3, Theorems 19.1 and 19.3] or [5, Theorems 35.4 and 16.13].

A.4.38 (Corollary to the martingale convergence theorem) Let $\mathcal{C}_1 \subset \mathcal{C}_2 \subset \cdots$ be an increasing sequence of sub-σ-algebras of $\mathcal{A}$ and set $\mathcal{C} = \sigma(\bigcup_n \mathcal{C}_n)$. Then $(E[f \mid \mathcal{C}_n], \mathcal{C}_n)_{n>0}$ is a uniformly integrable martingale and $\lim_{n\to\infty} E[f \mid \mathcal{C}_n] = E[f \mid \mathcal{C}]$ μ-a.e. for each $f \in L^1_\mu$, see [3, Theorems 19.3 and 19.5] or [5, Theorem 35.5].

A.4.39 (Corollary to the martingale convergence theorem) Let $(X, \mathcal{A}, \mu)$ be as before and let ν be another probability measure on $(X, \mathcal{A})$. Suppose that $\nu \ll \mu$. Let $f = \frac{d\nu}{d\mu}$ and $f_n = \frac{d\nu_{|\mathcal{C}_n}}{d\mu_{|\mathcal{C}_n}}$, $\mathcal{C}_n$ as before. Then $f_n = E[f \mid \mathcal{C}_n]$, $(f_n, \mathcal{C}_n)_{n>0}$ is a uniformly integrable martingale, and $f = \lim_{n\to\infty} f_n > 0$ ν-a.s. because $\nu\{f_n \le \epsilon\} \le \epsilon$ for all n.

Conditional probability distributions

Let $(X, \mathcal{B}, \mu)$ be a probability space and let $\mathcal{C}$ be a sub-σ-algebra of $\mathcal{B}$.

A.4.40 A *conditional probability distribution* for $\mu(\,.\mid \mathcal{C})$ is a family $(\mu_x \mid x \in X)$ of probability measures on $(X, \mathcal{B})$ such that

1. the functions $x \mapsto \mu_x(A)$ are $\mathcal{C}$-measurable for each $A \in \mathcal{B}$, and
2. $\mu(A) = \int \mu_x(A)\, d\mu(x)$ for all $A \in \mathcal{B}$.

A.4.41 Suppose X is a complete, separable metric space with Borel σ-algebra $\mathcal{B}$, μ is a Borel measure on X, and $\mathcal{C}$ is a sub-σ-algebra of $\mathcal{B}$. Then there is a conditional probability distribution $(\mu_x \mid x \in X)$ for $\mu(\,.\mid \mathcal{C})$. If $(\tilde{\mu}_x \mid x \in X)$ is another conditional probability distribution for $\mu(\,.\mid \mathcal{C})$, then $\mu_x = \tilde{\mu}_x$ for μ-a.e. x. (See [3, Theorems 44.3 and 44.2] or [5, Theorem 33.3].)

A.4.42 Let $(X, \mathcal{A}, \mu)$ and $\mathcal{C}$ be as before. As each bounded measurable function ψ on X can be uniformly approximated by elementary functions, the map $x \mapsto \int \psi\, d\mu_x$ is measurable for such ψ.

L^p-spaces

Let $(X, \mathcal{B}, \mu)$ be a probability space.

A.4.43 For $1 < p < \infty$ denote by L^p_μ the set of all measurable $f : X \to [-\infty, \infty]$ such that $\|f\|_p := (\int |f|^p\, d\mu)^{\frac{1}{p}} < \infty$. L^p_μ is a linear space and $\|\,.\,\|_p$ is a pseudonorm on L^p_μ, i.e., it is positive homogeneous and satisfies the triangle inequality. As $\|f\|_p = 0$ for each f with $f(x) = 0$ μ-a.s., $\|\,.\,\|_p$ is not a norm in general. The space of $\|\,.\,\|_p$-equivalence classes of functions in L^p_μ, however, is a Banach space, see [2, §15] or [50, Theorem 3.11].

Only at a few places in this book do we need to consider L^2_μ as a Banach space of equivalence classes of complex-valued functions. Observe that L^2_μ is even a Hilbert space where the norm $\|\,.\,\|_2$ comes from the inner product $\langle f, g\rangle = \int fg\, d\mu$, see [2, §15] or [50, Example 4.5]. This is the setting for the next two items.

A.4.44 (Orthogonal complements) For a linear subspace N of L^2_μ let $N^\perp := \{h \in L^2_\mu : \langle f, h\rangle = 0\ \forall f \in N\}$. Then $N^\perp$ is a closed linear subspace and as

$\mathrm{clos}(N) \oplus N^\perp = L^2_\mu$, $N^\perp$ is called the *orthogonal complement* of N, see [50, 4.9-11]. Observe that $N^\perp = \mathrm{clos}(N)^\perp$.

A.4.45 (Fourier series) Let $(X, \mathcal{B}, \mu)$ be the unit circle (considered as a subset of the complex plane) endowed with its Borel σ-algebra $\mathcal{B}$ and normalized one-dimensional Lebesgue measure μ. Let L^2_μ be the Hilbert space of equivalence classes of complex-valued square integrable functions on X. Then each $f \in L^2_\mu$ can be represented unambiguously as a *Fourier series* $f(z) = \sum_{n\in\mathbb{Z}} f_n \cdot z^n$ (convergence w.r.t. $\|\,.\,\|_2$) where $f_n \in \mathbb{C}$ and $\sum_{n\in\mathbb{Z}} |f_n|^2 = \|f\|_2^2$ [50, 4.26].

A.5 Making discontinuous mappings continuous

Consider a map $T : [0,1] \to [0,1]$ which is continuous except at a finite number of points $0 = a_0 < a_1 < \ldots < a_N = 1$ and monotone on each of the intervals (a_{i-1}, a_i). In particular, the one-sided limits of T at the points a_i exist. By T^n we denote the n-th iterate of T.

Even if T is not continuous at some of the a_i, the following trick turns it into a continuous map. Denote by

$$Y := \{x \in [0,1] \mid \exists k \geq 0\, \exists i \in \{1,\ldots,N-1\} : T^k x = a_i\}$$

the set of all discontinuities of all iterates T^k. We define X as the set that emerges from $[0,1]$ by replacing all points $y \in Y$ by a pair of points $y^- < y^+$. The natural order relation on $[0,1]$ carries over to X and is completed there by the additional requirement $y^- < y^+$. With this order relation X is a totally ordered, order complete set. The order topology on X is the topology generated by the open order intervals. It is a compact topology for X, see [7, Theorem X.7.12], and as the order intervals with endpoints in the countable set $(\mathbb{Q} \cap [0,1]) \cup \{y^-, y^+ : y \in Y\}$ form a basis of the topology, X is in fact a compact metrizable space. Denote by $\pi : X \to [0,1]$ the canonical map identifying pairs of points y^- and y^+. As T has one-sided limits at all points of $[0,1]$, it has a continuous extension to a transformation on $\tilde{T} : X \to X$ and $T \circ \pi = \pi \circ \tilde{T}$.

Let $\Sigma = \{1,\ldots,N\}$, $\Omega = \Sigma^{\mathbb{Z}_+}$, and define $\Phi : X \to \Omega$ by $(\Phi x)_k = j$ if $a^+_{j-1} \leq \tilde{T}^k x \leq a^-_j$. If T is strictly expanding on each interval (a_{j-1}, a_j) as, e.g., the skew tent map from Example 1.4.5, then similar arguments to the ones given in that example show that Φ is a homeomorphism from X onto Ω.

Given a subset $A \subseteq [0,1]$ denote by $\tilde{A} \subseteq X$ that set which arises from A by doubling all points in $A \cap Y$. As the open intervals in $[0,1]$ generate the topology on $[0,1]$, as the open order intervals in X generate the topology on X, and as corresponding intervals $I \subseteq [0,1]$ and $\tilde{I} \subseteq X$ with $\pi\tilde{I} = I$ differ by at most countably many points, the family

$$\{\tilde{A} : \pi\tilde{A} \subseteq [0,1] \text{ is a Borel set}\}$$

is exactly the Borel σ-algebra of X. Suppose now that μ is a non-atomic measure on $[0,1]$. As $\mu(Y)=0$, $\tilde{\mu}(\tilde{A}) := \mu(\pi\tilde{A})$ defines a non-atomic measure on X. Conversely, any non-atomic measure $\tilde{\mu}$ on X gives rise to a non-atomic measure μ on $[0,1]$ by $\mu(A) := \tilde{\mu}(\pi^{-1}A)$. Therefore, from a measure theoretic point of view, the maps T and $\tilde{T}$ are practically equivalent.

If an additional function $\psi : [0,1] \to \mathbb{R}$ is given which has at most finitely many discontinuities and which has one-sided limits everywhere, one can add the discontinuities of ψ to the set $\{a_0, \ldots, a_N\}$ and apply the above construction to this enlarged set. In this way one obtains a space X to which ψ can also be continuously extended as a function $\tilde{\psi} : X \to \mathbb{R}$.

References

[1] M. Barnsley, *Fractals Everywhere*, Second Edition, Academic Press Professional (1993).

[2] H. Bauer: *Maß- und Integrationstheorie*, de Gruyter Lehrbuch (1990).

[3] H. Bauer: *Wahrscheinlichkeitstheorie*, de Gruyter Lehrbuch (1991).

[4] T. Bedford, M. Keane, C. Series (eds.): *Ergodic Theory, Symbolic Dynamics and Hyperbolic Spaces*, Oxford University Press (1991).

[5] P. Billingsley, *Probability and Measure*, Second Edition, John Wiley & Sons (1986).

[6] R. Bowen: *Equilibrium States and the Ergodic Theory of Anosov Diffeomorphisms*, Lecture Notes in Mathematics, Vol. 470, Springer Verlag (1975).

[7] Garrett Birkhoff:[1] *Lattice Theory*, A.M.S. Colloquium Publications **25**.

[8] L. Boltzmann, Über die Eigenschaften monozyklischer und anderer damit verwandter Systeme, *Zeitschrift f. Reine u. Angew. Mathematik (Crelles Journal)* **98** (1884), 68–94.

[9] H. Bruin, G. Keller, Equilibrium states for S-unimodal maps, to appear in *Ergod. Th. & Dynam. Sys.*

[10] R. Burton, G. Keller, Stationary measures for randomly chosen maps, *J. Theor. Prob.* **6** (1993), 1–16.

[11] J. Buzzi, *Entropie et représentation markovienne des applications régulières sur l'intervalle*, Thesis, Paris (1995).

[12] D. Capocaccia: A definition of Gibbs state for a compact set with $\mathbb{Z}^\nu$ action, *Commun. Math. Phys.* **48**, 85–88 (1976).

[1] Garrett Birkhoff is not to be confused with George D. Birkhoff who first proved the ergodic theorem.

[13] I.P. Cornfeld, S.V. Fomin, Ya.G. Sinai: *Ergodic Theory*, Springer Verlag (1982).

[14] M. Denker, Ch. Grillenberger, K. Sigmund: *Ergodic Theory on Compact Spaces*, Lecture Notes in Mathematics, Vol. 527, Springer Verlag (1976).

[15] M. Denker, G. Keller, M. Urbanski: On the uniqueness of equilibrium states for piecewise monotone mappings, *Studia Mathematica* **97** (1990), 27–36.

[16] N. Dunford, J.T. Schwartz, *Linear Operators, Part I*, J. Wiley & Sons (1958).

[17] R. Durrett, An introduction to infinite particle systems, *Stoch. Proc. Appl.* **11** (1981), 109–150.

[18] P. and T. Ehrenfest: Begriffliche Grundlagen der statistischen Auffassung in der Mechanik, *Enzycl. d. Mathem. Wiss.* IV, 2, II, Heft 6 (1912), 3–90. English translation: *The Conceptual Foundations of the Statistical Approach in Mechanics*, Cornell University Press (1959).

[19] K. Falconer, *Techniques in Fractal Geometry*, J. Wiley & Sons (1997).

[20] P. Gänssler, W. Stute, *Wahrscheinlichkeitstheorie*, Springer Verlag (1977).

[21] D. Gatzouras, Y. Peres: The variational principle for the Hausdorff dimension: a survey, in: [48], pp. 113–126.

[22] C. F. Gauss, Mathematisches Tagebuch 1796–1814 (Note 113), in: *Carl Friedrich Gauss, Werke*, Band X/1, 483–574 (1917).

[23] H.-O. Georgii: *Gibbs Measures and Phase Transitions*, de Gruyter Studies in Mathematics (1988).

[24] J.R. Giles: *Convex Analysis with Applications in Differentiation of Convex Functions*, Research Notes in Mathematics, Vol. 58, Pitman (1982).

[25] R.M. Gray: *Probability, Random Processes, and Ergodic Properties*, Springer Verlag (1988).

[26] R.B. Griffiths: Peierls' proof of spontaneous magnetization in a two-dimensional Ising ferromagnet, *Phys. Rev.* **136 A** (1964), 437–439.

[27] B.M. Gurevich: Topological entropy of denumerable Markov chains, *Soviet. Math. Doklady* **10** (1969), 911–915.

[28] N.T.A. Haydn, D. Ruelle: Equivalence of Gibbs and equilibrium states for homeomorphisms satisfying expansiveness and specification, *Commun. Math. Phys.* **148** (1992), 155–167.

[29] F. Hofbauer: Examples for the nonuniqueness of the equilibrium state, *Transactions of the Amer. Math. Soc.* **228** (1977), 223–241.

[30] Y. Kamae, A simple proof of the ergodic theorem using nonstandard analysis, *Israel J. Math.* **42** (1982), 284–290.

[31] A. Katok, B. Hasselblatt, *Introduction to the Modern Theory of Dynamical Systems*, Cambridge University Press (1995).

[32] Y. Katznelson, B. Weiss, A simple proof of some ergodic theorems, *Israel J. Math.* **42** (291–296).

[33] G. Keller: Lifting measures to Markov extensions, *Monatsh. Math.* **108** (1989), 183–200.

[34] U. Krengel: *Ergodic Theorems*, de Gruyter Studies in Mathematics (1985).

[35] W. Krieger: On the uniqueness of equilibrium states, *Math. Sys. Theory* **8** (1974), 97–104.

[36] F. Ledrappier: Principe variationnel et systèmes dynamiques symboliques, *Z. Wahrscheinlichkeitstheorie verw. Gebiete* **30** (1974), 185–202.

[37] R. Mañé: *Ergodic Theory and Differentiable Dynamics*, Ergebnisse der Mathematik und ihrer Grenzgebiete, 3. Folge, Band 8, Springer Verlag (1987).

[38] N.F.G. Martin, J.W. England, *Mathematical Theory of Entropy*, Encyclopedia of Mathematics and its Applications, Vol. 12, Addison-Wesley (1981).

[39] M. Mathieu, On the origin of the notion 'Ergodic Theory', *Expo. Math.* **6** (1988), 373–377.

[40] W. de Melo, S. van Strien: *One-Dimensional Dynamics*, Springer Verlag (1993).

[41] M. Misiurewicz: A short proof of the variational principle for a $\mathbb{Z}_+^N$-action on a compact space, *Bulletin de l'Académie Polonaise des Sciences* **24** (1976), 1069–1075.

[42] J. Moulin Ollagnier: *Ergodic Theory and Statistical Mechanics*, Lecture Notes in Mathematics, Vol. 1115, Springer Verlag (1985).

[43] W. Parry: *Entropy and Generators in Ergodic Theory*, Benjamin (1969).

[44] R. Peierls: On Ising's model of ferromagnetism, *Proc. Cambr. Phil. Soc.* **32** (1936), 477–481.

[45] Ya.B. Pesin: Dimension type characteristics for invariant sets of dynamical systems, *Russian Math. Surveys* **43** (1988), 111–151.

[46] K. Petersen, *Ergodic Theory*, Cambridge University Press (1983).

[47] M. Pollicott, *Lectures on Ergodic Theory and Pesin Theory on Compact Manifolds*, London Math. Soc. Lecture Note Series, Vol. 180, Cambridge University Press (1993).

[48] M. Pollicott, K. Schmidt (eds.): *Ergodic Theory of $\mathbb{Z}^d$-Actions*, London Math. Soc. Lecture Note Series, Vol. 228, Cambridge University Press (1996).

[49] B. Prum, J.C. Fort, *Stochastic Processes on a Lattice and Gibbs Measures*, Mathematical Physics Studies, Vol. 11, Kluwer Academic Publishers (1991).

[50] W. Rudin, *Real and Complex Analysis*, McGraw-Hill (1970).

[51] D.J. Rudolph: *Fundamentals of Measurable Dynamics*, Clarendon Press (1990).

[52] D. Ruelle: *Thermodynamic Formalism*, Encyclopedia of Mathematics and its Applications, Vol. 5, Addison-Wesley (1978).

[53] D. Ruelle: Thermodynamic formalism for maps satisfying positive expansiveness and specification, *Nonlinearity* **5** (1992), 1223–1236.

[54] K. Schmidt: *Dynamical Systems of Algebraic Origin*, Progress in Mathematics, Vol. 128, Birkhäuser (1995).

[55] F. Schweiger: *Ergodic Theory of Fibred Systems and Metric Number Theory*, Oxford Science Publications, Clarendon Press (1995).

[56] E. Seneta, *Non-negative Matrices*, George Allen & Unwin (1973).

[57] Y. Takahashi: Entropy functional (free energy) for dynamical systems and their random perturbations, in: *Taniguchi Symp. SA, Katata* (1982), 437–467.

[58] J. Väisälä, *Lectures on n-dimensional Quasiconformal Mappings*, Lecture Notes in Mathematics, Vol. 229, Springer Verlag (1971).

[59] M. Vuorinen, *Conformal Geometry and Quasiregular Mappings*, Lecture Notes in Mathematics, Vol. 1319, Springer Verlag (1988).

[60] P. Walters: *An Introduction to Ergodic Theory*, Springer Verlag (1982).

[61] P. Walters: A variational principle for the pressure of continuous transformations, *Amer. J. Math.* **97** (1976), 937–971.

[62] L.S. Young: Large deviations in dynamical systems, *Transactions of the Amer. Math. Soc.* **318** (1990), 525–543.

List of special notations

In this list of special notations and in the following index, bold page numbers refer to the most important occurences of the entries.

Index